Geometry
DeMYSTiFieD®

DeMYSTiFieD® Series

Accounting Demystified
Advanced Calculus Demystified
Advanced Physics Demystified
Advanced Statistics Demystified
Algebra Demystified
Alternative Energy Demystified
Anatomy Demystified
asp.net 2.0 Demystified
Astronomy Demystified
Audio Demystified
Biology Demystified
Biotechnology Demystified
Business Calculus Demystified
Business Math Demystified
Business Statistics Demystified
C++ Demystified
Calculus Demystified
Chemistry Demystified
Circuit Analysis Demystified
College Algebra Demystified
Corporate Finance Demystified
Databases Demystified
Data Structures Demystified
Differential Equations Demystified
Digital Electronics Demystified
Earth Science Demystified
Electricity Demystified
Electronics Demystified
Engineering Statistics Demystified
Environmental Science Demystified
Everyday Math Demystified
Fertility Demystified
Financial Planning Demystified
Forensics Demystified
French Demystified
Genetics Demystified
Geometry Demystified
German Demystified
Home Networking Demystified
Investing Demystified
Italian Demystified
Java Demystified
JavaScript Demystified
Lean Six Sigma Demystified
Linear Algebra Demystified

Logic Demystified
Macroeconomics Demystified
Management Accounting Demystified
Math Proofs Demystified
Math Word Problems Demystified
MATLAB® Demystified
Medical Billing and Coding Demystified
Medical Terminology Demystified
Meteorology Demystified
Microbiology Demystified
Microeconomics Demystified
Nanotechnology Demystified
Nurse Management Demystified
OOP Demystified
Options Demystified
Organic Chemistry Demystified
Personal Computing Demystified
Pharmacology Demystified
Physics Demystified
Physiology Demystified
Pre-Algebra Demystified
Precalculus Demystified
Probability Demystified
Project Management Demystified
Psychology Demystified
Quality Management Demystified
Quantum Mechanics Demystified
Real Estate Math Demystified
Relativity Demystified
Robotics Demystified
Sales Management Demystified
Signals and Systems Demystified
Six Sigma Demystified
Spanish Demystified
sql Demystified
Statics and Dynamics Demystified
Statistics Demystified
Technical Analysis Demystified
Technical Math Demystified
Trigonometry Demystified
uml Demystified
Visual Basic 2005 Demystified
Visual C# 2005 Demystified
xml Demystified

Geometry
DeMYSTiFieD®

Stan Gibilisco

Second Edition

New York   Chicago   San Francisco   Lisbon   London   Madrid   Mexico City
Milan   New Delhi   San Juan   Seoul   Singapore   Sydney   Toronto

Cataloging-in-Publication Data is on file with the Library of Congress.

McGraw-Hill books are available at special quantity discounts to use as premiums and sales promotions, or for use in corporate training programs. To contact a representative please e-mail us at bulksales@mcgraw-hill.com.

Geometry DeMYSTiFieD®, Second Edition

2 3 4 5 6 7 8 9 0 DOC/DOC 1 9 8 7 6 5 4 3

ISBN 978-0-07-175626-6
MHID 0-07-175626-4

Sponsoring Editor
Judy Bass

Acquisitions Coordinator
Michael Mulcahy

Editing Supervisor
David E. Fogarty

Project Manager
Tania Andrabi,
Glyph International

Copy Editor
Manish Tiwari,
Glyph International

Proofreader
Richa Sodhi,
Glyph International

Production Supervisor
Pamela A. Pelton

Composition
Glyph International

Art Director, Cover
Jeff Weeks

Cover Illustration
Lance Lekander

To Samuel, Tony, and Tim
from Uncle Stan

About the Author

Stan Gibilisco, an electronics engineer, researcher, and mathematician, has authored multiple titles for the McGraw-Hill *Demystified* and *Know-It-All* series, along with numerous other technical books and dozens of magazine articles. His work has been published in several languages.

Contents

Acknowledgments

I extend thanks to my nephew Tim Boutelle, who helped me proofread the manuscript and offered suggestions from the viewpoint of the intended audience.

How to Use This Book

This book can help you learn basic geometry without taking a formal course. It can also serve as a supplemental text in a classroom, tutored, or homeschooling environment.

None of the mathematics in this book goes beyond the high-school level. If you need a "refresher," you can select from several *Demystified* books dedicated to mathematics topics. If you want to build yourself a "rock-solid" mathematics foundation before you start this course, I recommend that you go through *Pre-Algebra Demystified*, *Algebra Demystified*, and *Algebra Know-It-All*.

This book contains abundant multiple-choice questions written in standardized test format. You'll find an "open-book" quiz at the end of every chapter. You may (and should) refer to the chapter texts when taking these quizzes. Write down your answers, and then give your list of answers to a friend. Have your friend tell you your score, but not which questions you missed. The correct answers appear in the back of the book. Stick with a chapter until you get most of the quiz answers correct.

Two major sections constitute this course. Each section ends with a multiple-choice test. Take these tests when you're done with the respective sections and have taken all the chapter quizzes. Don't look back at the text when taking the section tests. They're easier than the chapter-ending quizzes, and they don't require you to memorize trivial things. A satisfactory score is three-quarters correct. Answers appear in the back of the book.

The course concludes with a 100-question final exam. Take it when you've finished all the sections, all the section tests, and all of the chapter quizzes. A satisfactory score is at least 75 percent correct answers.

With the section tests and the final exam, as with the quizzes, have a friend divulge your score without letting you know which questions you missed. That way, you won't subconsciously memorize the answers. You might want to take each test, and the final exam, two or three times. When you get a score that makes you happy, you can (and should) check to see where your strengths and weaknesses lie.

You won't find any proofs here. Instead of taking up a lot of space with theorem demonstrations, this course concentrates on fundamental facts and a diversity of topics found in few, if any, other introductory geometry texts. If you're interested in learning how to do proofs, I recommend *Math Proofs Demystified*. If you want to delve further into analytic geometry and vectors, I recommend *Pre-Calculus Know-It All*.

Strive to complete one chapter of this book every 10 days or 2 weeks. Don't rush, but don't go too slowly either. Proceed at a steady pace and keep it up. That way, you'll complete the course in a few months. (As much as we all wish otherwise, nothing can substitute for "good study habits.") When you're done with the course, you can use this book as a permanent reference.

I welcome your ideas and suggestions for future editions.

Stan Gibilisco

Geometry
DeMYSTiFieD®

Part I

Two Dimensions

1

Rules of the Game

The fundamental rules of geometry date back to the time of the ancient Egyptians and Greeks, who used geometry to calculate the diameter of the earth and the distance to the moon. These mathematicians employed the laws of *Euclidean geometry* (named after *Euclid of Alexandria*, a Greek mathematician who lived around the third century B.C.). Two-dimensional Euclidean geometry, also called *plane geometry*, involves points, lines, and shapes confined to flat surfaces.

CHAPTER OBJECTIVES

In this chapter, you will

- Envision "mathematically perfect" points and straight lines.
- Break lines up into rays and segments.
- Define angles and distances.
- Measure and compare angles.
- Add and subtract angles.
- Learn how lines and angles relate.

Points and Lines

In plane geometry we regard certain concepts as intuitively obvious without the need for formal definitions. We call these "mathematically perfect things" *elementary objects*: the *point*, the *line*, and the *plane*. We can imagine a point as an infinitely tiny sphere having height, width, and depth all equal to zero, but nevertheless possessing a specific location. We can think of a line as an infinitely thin, perfectly straight, infinitely long wire or thread. We can imagine a plane as an infinitely thin, perfectly flat surface having an infinite expanse.

Naming Points and Lines

Geometers name points and lines using uppercase, italicized letters of the alphabet. The most common name for a point is P (for "point"), and the most common name for a line is L (for "line"). If we have multiple points in a situation, we can use the letters P, Q, R, S, and so on all the way to Z if needed. If two or more lines exist in a scenario, we can use the letters immediately following L, all the way up to N. (We should try to avoid using the uppercase O because it looks a lot like the numeral 0!) Alternatively, we can use numeric subscripts with the uppercase, italic letters P and L, naming points P_1, P_2, P_3, ..., P_n, and naming lines L_1, L_2, L_3, ..., L_n (where n represents an arbitrary positive whole number that's as large as we need).

Two-Point Principle

Suppose that P and Q represent different geometric points. These points define *one and only one* line L (i.e., a *unique* line L). The following two statements always hold true in a situation like this, as shown in Fig. 1-1:

- Points P and Q lie on a common line L.
- Line L is the only line on which both points lie.

Distance Notation

We can symbolize the distance between any two points P and Q, as we express it going from P toward Q along the straight line connecting them, by writing

FIGURE 1-1 • The two-point principle.

PQ. Units of measurement such as meters, feet, millimeters, inches, miles, or kilometers have no relevance in pure mathematics, but they're important in physics and engineering. As an alternative notation, we can use a lowercase letter such as *d* to represent the distance between two points.

Line Segments

The portion of a line between two different points *P* and *Q* constitutes a *line segment*. We call the points *P* and *Q* the *end points*. A line segment can theoretically include both of the end points, only one of them, or neither of them. Therefore, three possibilities exist, as follows:

- If a line segment contains both end points, we call it a *closed line segment*. We indicate the fact that the end points are included by drawing them both as solid black dots.
- If a line segment contains one of the end points but not the other, we call it a *half-open line segment*. We draw the included end point as a solid black dot and the excluded end point as a small open circle.
- If a line segment contains neither end point, we call it an *open line segment*. We draw both end points as small open circles.

TIP *Any particular line segment has the same length, regardless of whether it's closed, half-open, or open. Adding or taking away a single point makes no difference, mathematically, in the length, because points have zero size in all dimensions!*

Rays (Half Lines)

Sometimes, mathematicians talk about the portion of a geometric line that lies "on one side" of a certain point. In the situation of Fig. 1-1, imagine the set of points that starts at *P*, then passes through *Q*, and extends onward past *Q* forever. We call the resulting object a *ray* or *half line*. The ray defined by *P* and *Q* might include the end point *P*, in which case we have a *closed-ended ray*. If we leave the end point out, we get an *open-ended ray*. Either way, we say that the ray or half line *begins* or *originates* at point *P*.

Midpoint Principle

Imagine a line segment connecting two points *P* and *R*. There exists one and only one point *Q* on the line segment such that *PQ* = *QR*, as shown in Fig. 1-2.

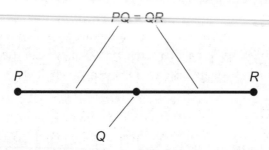

FIGURE 1-2 · The midpoint principle.

PROBLEM 1-1

Suppose that, in the scenario of Fig. 1-2, we find the midpoint Q_2 between P and Q, then the midpoint Q_3 between P and Q_2, then the midpoint Q_4 between P and Q_3, and so on. In mathematical language, we say that we keep finding midpoints $Q_{(n+1)}$ between P and Q_n, where n represents a positive whole number. How long can we continue this process?

SOLUTION

The process can continue forever. In theoretical geometry, no limit exists as to the number of times we can cut a line segment in half, because a line segment contains infinitely many points.

PROBLEM 1-2

Imagine a line segment with end points P and Q. What's the difference between the distance PQ and the distance QP?

SOLUTION

If we consider distance without paying attention to the direction in which we express or measure it, then $PQ = QP$. But if the direction does make a difference to us, we can define $PQ = -QP$. Then we use the term *displacement* instead of *direction*.

In geometry diagrams, we can specify displacements (instead of simple distances) if we want to induce our readers to move their eyes from right to left instead of from left to right, or from bottom to top rather than from top to bottom.

Angles and Distances

When two lines intersect, we get four distinct *angles* at the point of intersection. In most cases, we'll find that two of the angles are "sharp" and two are "dull." If all four of the angles happen to turn out identical, then they all constitute *right angles*, and we say that the lines run *perpendicular, orthogonal,* or *normal* to each other at the point of intersection. We can also define an angle using three points connected by two line segments; the angle appears at the point where the line segments meet.

Measuring Angles

To express the extent or measure of an angle, we can use either of two units: the *degree* and the *radian*. The degree (°) is the unit familiar to lay people, while the radian is more often used by mathematicians and engineers.

One degree (1°) equals 1/360 of a full circle. Therefore, 90° represents 1/4 of a circle, 180° represents a half circle, 270° represents 3/4 of a circle, and 360° represents a full circle. A *right angle* has a measure of 90°, an *acute angle* has a measure of more than 0° but less than 90°, and an *obtuse angle* has an angle more than 90° but less than 180°. A *straight angle* has a measure of 180°. A *reflex angle* has a measure of more than 180° but less than 360°.

We can define the radian (rad) as follows. Imagine two rays emanating outward from the center point of a circle. Each of the two rays intersects the circle at a point; call these points P and Q. Suppose that the distance between P and Q, as expressed along the arc of the circle, equals the radius of the circle. Then the measure of the angle between the rays equals 1 radian (1 rad).

A full circle contains 2π rad, where π (the lowercase Greek letter pi, pronounced "pie") stands for the ratio of a circle's circumference to its diameter. The value of π is approximately 3.14159265359, often rounded off to 3.14159 or 3.14. A right angle has a measure of $\pi/2$ rad, an acute angle has a measure of more than 0 rad but less than $\pi/2$ rad, and an obtuse angle has an angle more than $\pi/2$ rad but less than π rad. A straight angle has a measure of π rad, and a reflex angle has a measure larger than π rad but less than 2π rad.

TIP *Mathematicians often delete the unit reference when they express or write about angles in radians. Therefore, instead of "$\pi/3$ rad," you might encounter an angle denoted as "$\pi/3$." Whenever you see a reference to an angle and no unit goes along with it, you can assume that the author is working with radians.*

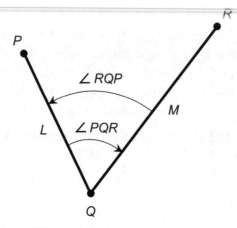

FIGURE 1-3 · Angle notation.

Angle Notation

Imagine that *P*, *Q*, and *R* represent three distinct points. Let *L* represent the line segment connecting *P* and *Q*, and let *M* represent the line segment connecting *R* and *Q*. We can denote the angle between *L* and *M*, as measured at point *Q* in the plane defined by the three points, by writing ∠PQR or ∠RQP as shown in Fig. 1-3.

If we want to specify the *rotational sense* of the angle (either *counterclockwise* or *clockwise*), then ∠RQP indicates the angle as we turn counterclockwise from *M* to *L*, and ∠PQR indicates the angle as we turn clockwise from *L* to *M*. We consider counterclockwise-going angles as having positive values and clockwise-going angles as having negative values.

In the situation of Fig. 1-3, ∠RQP is positive while ∠PQR is negative. If we make an approximate guess as to the measures of the angles in Fig. 1-3, we might say that ∠RQP ≈ +60° while ∠PQR ≈ –60°. The "wavy" equals sign translates literally to the phrase "approximately equals" or the phrase "is approximately equal to."

Still Struggling

Rotational sense doesn't matter in basic geometry. However, it does matter when we work in *coordinate geometry* (geometry involving graphs). We'll get into coordinate geometry, also known as *analytic geometry*, later in this book. For now, let's not worry about the rotational sense in which we express or measure an angle. We can consider all angles as having positive measures.

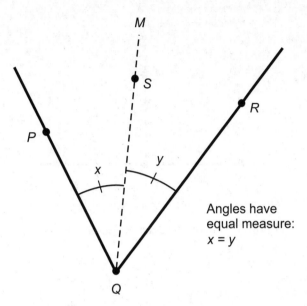

FIGURE 1-4 · The angle bisection principle.

Angle Bisection Principle

Consider an angle $\angle PQR$ measuring less than 180° and defined by three points P, Q, and R as shown in Fig. 1-4. There exists exactly one ray M that *bisects* (divides in half) the angle $\angle PQR$. If S represents any point on M other than point Q, then $\angle PQS = \angle SQR$. Every angle has *one and only one* ray that bisects it.

Perpendicular Principle

Consider a line L that passes through points P and Q. Let R represent a point that does not lie on L. There exists exactly one line M through point R, intersecting line L at some point S, such that M runs perpendicular to L (M and L intersect at a right angle) at point S. Figure 1-5 illustrates this situation.

Perpendicular Bisector Principle

Suppose that L represents a line segment connecting two points P and R. There exists one and only one line M that runs perpendicular to L and that intersects L at a point Q, such that the distance from P to Q equals the distance from Q to R. In other words, every line segment has exactly one *perpendicular bisector*. Figure 1-6 illustrates this situation.

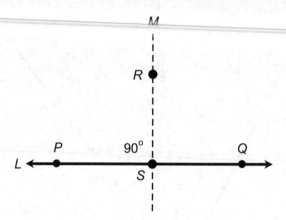

FIGURE 1-5 • The perpendicular principle.

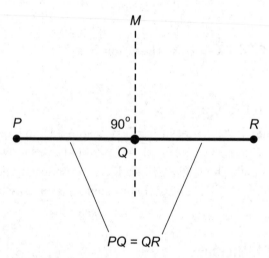

FIGURE 1-6 • The perpendicular bisector principle.

Distance Addition and Subtraction

Let P, Q, and R represent points on a line L, such that Q lies between P and R. The following equations hold concerning distances as measured along L (Fig. 1-7):

$$PQ + QR = PR$$
$$PR - PQ = QR$$
$$PR - QR = PQ$$

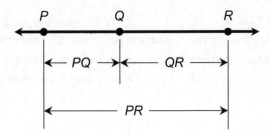

FIGURE 1-7 · Distance addition and subtraction.

Angular Addition and Subtraction

Suppose that P, Q, R, and S represent points that all lie in the same plane. In other words, all four points lie on a single, perfectly flat surface. Let Q represent the vertex of three angles $\angle PQR$, $\angle PQS$, and $\angle SQR$, with ray QS between rays QP and QR as shown in Fig. 1-8. The following equations hold concerning the angular measures:

$$\angle PQS + \angle SQR = \angle PQR$$
$$\angle PQR - \angle PQS = \angle SQR$$
$$\angle PQR - \angle SQR = \angle PQS$$

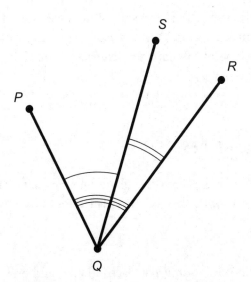

FIGURE 1-8 · Angular addition and subtraction.

PROBLEM 1-3

Examine Fig. 1-6 once again. Imagine some point *S*, other than point *Q*, that lies on line *M* (the perpendicular bisector of the line segment connecting *P* and *R*). What can you say about the lengths of line segments *PS* and *SR*?

SOLUTION

You can "streamline" the solutions to problems like this by making your own drawings. As the language gets more complicated (geometry problems can sometimes read like "legalese"), such drawings become increasingly helpful. With the aid of your own sketch, you should see that for every point *S* on line *M* (other than point *Q*), the distance *PS* exceeds the distance *PQ* (i.e., *PS* > *PQ*), and the distance *SR* exceeds the distance *QR* (i.e., *SR* > *QR*).

PROBLEM 1-4

Look again at Fig. 1-8. Suppose that you move point *S* either straight toward yourself (out of the page) or straight away from yourself (back behind the page), so *S* no longer lies in the same plane as points *P*, *Q*, and *R*. What can you say about the measures of ∠PQR, ∠PQS, and ∠SQR?

SOLUTION

In either of these situations, the sum of the measures of ∠PQS and ∠SQR exceeds the measure of ∠PQR, because the measures of ∠PQS and ∠SQR both increase if point *S* departs *perpendicularly* from the plane containing points *P*, *Q*, and *R*.

More about Lines and Angles

If we remain within a single geometric plane, lines and angles behave according to various rules. Some of the best-known principles follow.

Parallel Lines

We say that two lines run *parallel* to each other if and only if they lie in the same plane and they don't intersect at any point. Two line segments or rays run parallel to each other if and only if, when extended infinitely in both directions to form complete lines, those lines don't intersect at any point.

Complementary and Supplementary Angles

We say that two angles in the same plane constitute *complementary angles* (they "complement" each other) if and only if the sum of their measures equals 90° (π/2 rad). We say that two angles in the same plane constitute *supplementary angles* (they "supplement" each other) if and only if the sum of their measures equals 180° (π rad).

Adjacent Angles

Consider two lines L and M that intersect at a point P. Any two *adjacent angles* (i.e., any two angles that lie next to each other) between lines L and M are supplementary. We can illustrate this fact by drawing two intersecting lines and noting that pairs of adjacent angles always form a *straight angle*, that is, an angle of 180° (π rad) determined by the intersection point and either of the two lines.

Vertical Angles

Again consider two lines L and M that intersect at a point P. We call the opposing pairs of angles, denoted as x and y in Fig. 1-9, *vertical angles*. In any situation of this sort, the vertical angles have equal measure.

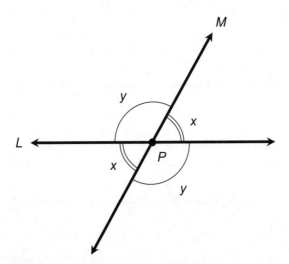

FIGURE 1-9 · Vertical angles between two intersecting lines.

Still Struggling

The term "vertical" to describe angles such as those shown in Fig. 1-9 baffles some people. They don't look "vertical," do they? We might do better to call such angles "opposite" or "opposing." But a long time ago, somebody decided that the term "vertical" was good enough, and no one has ever changed it.

Transversals and Interior Angles

Imagine two lines L and M that lie in the same plane. Let N represent a line that intersects L and M at points P and Q, respectively. We call line N a *transversal* to the lines L and M. In Fig. 1-10, the angles labeled x and z constitute a pair of *alternate interior angles*. The same holds true for the pair of angles labeled w and y.

When we confine our attention to a single geometric plane, pairs of alternate interior angles formed by a transversal line have equal measure if and only if the two lines crossed by the transversal run parallel to each other. The pairs of alternate interior angles *do not* have equal measure if and only if the two lines crossed by the transversal *do not* run parallel to each other.

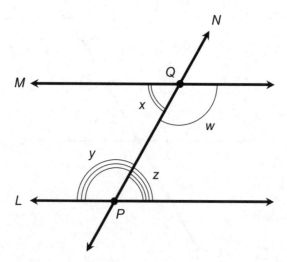

FIGURE 1-10 · Alternate interior angles formed by a transversal line.

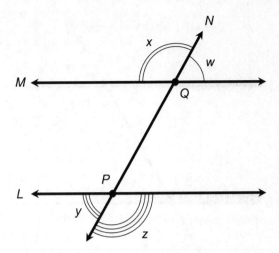

FIGURE 1-11 · Alternate exterior angles formed by a transversal line.

Transversals and Exterior Angles

Again, imagine two lines L and M that lie in the same plane, and that are both crossed by a transversal line N at points P and Q. In Fig. 1-11, the two angles labeled x and z are *alternate exterior angles*, so are the two angles labeled w and y.

Within a single geometric plane, pairs of alternate exterior angles formed by a transversal line have equal measure if and only if the two lines crossed by the transversal run parallel to each other. The pairs of alternate exterior angles *do not* have equal measure if and only if the two lines crossed by the transversal *do not* run parallel to each other.

Corresponding Angles

Now consider two lines L and M that lie in the same plane, and that also happen to run parallel to each other. Let N represent a transversal that intersects L and M at points P and Q, respectively. We've learned that in this special situation both pairs of alternate interior angles have equal measure, and both pairs of alternate exterior angles have equal measure. But we can say more! In the situation of Fig. 1-12, each of the four pairs of angles "facing in the same direction" constitutes *corresponding angles*, as follows:

- The two angles w correspond.
- The two angles x correspond.
- The two angles y correspond.
- The two angles z correspond.

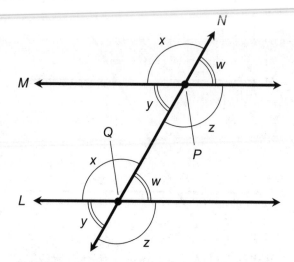

FIGURE 1-12 · Corresponding angles have equal measure if and only if a transversal crosses two parallel lines.

Whenever a transversal crosses two parallel lines, each individual pair of corresponding angles has equal measure.

Perpendicular Transversal

Given two parallel lines L and M along with a transversal N that crosses them both, we can be certain that N runs perpendicular to both L and M (i.e., N is a perpendicular transversal to the parallel lines L and M) if and only if any single pair of adjacent angles has equal measure.

Parallel Principle

Suppose that L represents a line and P represents a point that doesn't lie on L. In any situation of this sort, there exists one and only one line M through P, such that M runs parallel to L (Fig. 1-13). We call this fact the *parallel principle* or *parallel postulate*. It constitutes one of the most important postulates in Euclidean geometry.

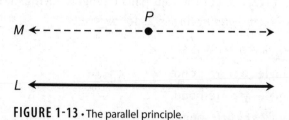

FIGURE 1-13 · The parallel principle.

TIP *In certain variants of geometry, the parallel postulate* does not *hold true. The denial of the parallel postulate forms the cornerstone of* non-Euclidean *geometry. We'll delve into that subject in Chap. 11.*

The Parallel Principle Repeated

Let L and M represent two different lines that lie in the same plane. Suppose that both L and M intersect a transversal line N and both L and M run perpendicular to N. Then lines L and M are parallel to each other (Fig. 1-14). We can call this fact the *principle of mutual perpendicularity*. In Fig. 1-14, we illustrate the fact that two lines run perpendicular to each other by marking the intersection point with a small square. Geometers commonly use this trick to show that lines, line segments, or rays intersect at right angles.

PROBLEM 1-5

Imagine that you stand on the edge of a highway. The road is perfectly straight and flat, and the pavement is 20 meters wide everywhere. Suppose that you lay a string across the road so that it intersects one edge of the pavement at a 70° angle, measured with respect to the edge itself. If you stretch the string out perfectly straight and then you reel out enough string so it crosses the other edge of the road, at what angle will the string intersect the other edge of the pavement, measured relative to that edge? At what angle will the string intersect the centerline of the road, measured relative to the centerline?

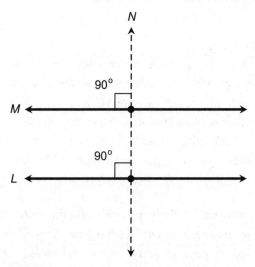

FIGURE 1-14 • Mutual perpendicularity

SOLUTION

 This problem involves a double case of alternate interior angles, illustrated in Fig. 1-10. Alternatively, you can employ the principle for corresponding angles (Fig. 1-12). The edges of the pavement run parallel to each other, and both edges run parallel to the centerline. Therefore, the string will intersect the other edge of the road at a 70° angle; it will also cross the centerline at a 70° angle. Note that these angles are expressed between the string and the pavement edges and centerline themselves, not with respect to the lines that run normal to the pavement edge or the centerline (as is often done in physics).

Still Struggling

In the foregoing solution, we specify the smaller of two intersection angles between the string and the road edges, and between the string and the centerline. We could also use the larger angle of 110°, which represents the supplement of 70°.

PROBLEM **1-6**

What are the measures of the angles described in Problem 1-5 and its solution with respect to normals to the pavement edges and centerline?

SOLUTION

A normal to any line always subtends an angle of 90° relative to that line. Therefore, the string will cross both edges of the pavement at an angle of 90° – 70°, or 20°, relative to the normal. We know this fact from the principle of angle addition and subtraction. The string will also cross the centerline at an angle of 20° with respect to the normal. We get the same 20° result if we use the larger angle, because 110° – 90° = 20°.

TIP *Obviously, no one should conduct experiments like those of Problems 1-5 and 1-6 on real roads. If you want to check out the foregoing facts on a big scale, draw "fake roads" with chalk on your own driveway, or draw lines in the sand at the beach! Leave irresponsible road experiments to wild animals!*

QUIZ

Refer to the text in this chapter if necessary. A good score is eight correct. Answers are in the back of the book.

1. What distinction, if any, exists between the meanings of *distance* and *displacement*?
   A. None! The two terms have identical meanings.
   B. Displacement refers to distance in a specified direction.
   C. Distance refers to displacement in a specified direction.
   D. Displacement refers to the speed of physical motion from one point to another, while distance refers only to the separation between two points.

2. An angle having a measure of 315° constitutes
   A. 3/8 of a full circle.
   B. 5/8 of a full circle.
   C. 7/8 of a full circle.
   D. more than full circle.

3. An angle having a measure of $\pi/3$ constitutes
   A. an acute angle.
   B. an obtuse angle.
   C. a reflex angle.
   D. more than a full circle.

4. Imagine that you encounter a straight, infinitely long line. You choose a point that doesn't lie on that line. (Any point will do.) Then you attempt to draw a new line that runs through the point you've just chosen, and that also runs through the original line at a right angle. How many such lines can you find?
   A. None
   B. One
   C. Two
   D. Infinitely many

5. Imagine that you encounter a straight line segment having finite length. You attempt to draw an infinitely long, straight line that passes through the original line segment at a right angle, and that also divides the original line segment into two identical halves. How many such lines can you find?
   A. None
   B. One
   C. Two
   D. Infinitely many

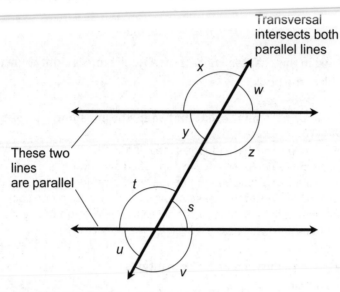

FIGURE 1-15 · Illustration for Quiz Questions 6 and 7.

6. Figure 1-15 illustrates a transversal line that passes through two parallel lines. The two intersection points produce eight angles, labeled as shown. Which of the following pairs of angles *do not* necessarily have equal measure?

A. *w* and *s*

B. *w* and *u*

C. *v* and *y*

D. *v* and *z*

7. Suppose that, in the situation of Fig. 1-15, we "adjust" the transversal line so that angles *s* and *v* have equal measure. In that case, we know that the transversal line

A. runs perpendicular to both parallel lines.

B. can't possibly run perpendicular to either of the parallel lines.

C. can run perpendicular to only one of the parallel lines.

D. lies outside of the plane formed by the parallel lines.

8. Between a pair of intersecting lines, the sum of the measures of two adjacent angles is always

A. less than π.

B. equal to π.

C. more than π.

D. equal to 2π.

9. Imagine that you encounter a straight, infinitely long line. You choose a point that doesn't lie on that line. (Any point will do.) Then you attempt to draw a new

line that runs through the point you've just chosen, and that also intersects the original line. How many such lines can you find?

A. None

B. One

C. Two

D. Infinitely many

10. **When we have a transversal line that crosses two parallel lines, we get eight angles at the two points where the three lines intersect. Which of the following types of angle pairs always have equal measure?**

A. Corresponding angles

B. Alternate interior angles

C. Alternate exterior angles

D. All of the above

2

Triangles

If you ever took a course in plane geometry, you remember triangles. Do you recall dragging your mind through formal proofs about them? You won't have to scrutinize any proofs here, but you should know some basic facts about triangles. Even if you've never worked with triangles before, you should find the information in this chapter easy to grasp.

CHAPTER OBJECTIVES

In this chapter, you will

- Define, name, and analyze triangles.
- Learn about similarity and congruence.
- Learn how to uniquely determine the sides and angles of any triangle.
- Classify triangles according to shape.
- Discover the Theorem of Pythagoras.
- Calculate the perimeters and interior areas of triangles.

Triangle Definitions

In mathematics, we should always have a firm grasp of what we're talking about, without any vagueness or ambiguity. That's why we need formal definitions for everything except elementary concepts such as the point, line, and plane.

What's a Triangle?

A *triangle* comprises three line segments, joined pairwise at their end points, and including those end points. The three points must not be *collinear*. That is, they must not all lie on the same straight line. For our purposes, we assume that the universe in which we define the triangle is Euclidean or "flat," not "curved" like the surface of the earth or "warped" like the space around a black hole. In such an ideal universe, we can always define the shortest distance between two different points by finding the straight line segment connecting those two points and then measuring the length of that segment.

Vertices

Figure 2-1 shows three points called *A*, *B*, and *C*, connected by line segments to form a triangle. We call these points the *vertices* of the triangle. We can use almost any uppercase, italicized alphabetic letter to denote the vertices of a triangle. The letters *P*, *Q*, and *R* are common alternatives to *A*, *B*, and *C*.

Naming

We can call the triangle in Fig. 2-1 "triangle *ABC*." Geometers sometimes write a little triangle symbol (Δ) in place of the word "triangle." This symbol is the uppercase Greek letter delta. Figure 2-1 therefore portrays an arbitrary Δ*ABC*.

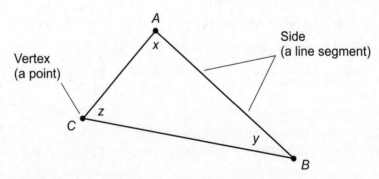

FIGURE 2-1 · Vertices, sides, and angles of a triangle.

We can list the vertices in any other order if we want, so we can call the triangle of Fig. 2-1 by any of the following names:

$$\triangle ABC$$

$$\triangle BCA$$

$$\triangle CAB$$

$$\triangle CBA$$

$$\triangle BAC$$

$$\triangle ACB$$

Sides

We can name the sides of the triangle in Fig. 2-1 according to their end points. Thus, $\triangle ABC$ has three sides: line segment AB, line segment BC, and line segment CA. There are other ways of naming the sides; as long as we don't confuse anybody, we can call them anything we want.

Interior Angles

Each vertex of a triangle corresponds to a specific *interior angle*, which always measures more than 0° (0 rad) but less than 180° (π rad). In Fig. 2-1, we denote the interior angles with the lowercase italicized English letters $x, y,$ and z. Some mathematicians prefer to use italic lowercase Greek letters to symbolize angles. Theta (pronounced THAY-tuh) is a popular choice. It looks like a leaning numeral zero with a dash across it (θ). Subscripts can help us denote the interior angles of a triangle, for example, θ_a, θ_b, and θ_c for the interior angles at vertices A, B, and C, respectively. As with the sides, we can give the angles any names we want, as long as each angle gets its own name.

Similar Triangles

Two triangles are *directly similar* if and only if they have the same proportions in the same rotational sense, that is, as we go around them both in the same direction (either clockwise or counterclockwise). Therefore, one triangle constitutes an enlarged, reduced, and/or rotated copy of the other. They can also have identical size, shape, and orientation by coincidence.

Figure 2-2 shows some examples of directly similar triangles. If you take any one of the triangles, enlarge it or reduce it uniformly, and rotate it (if necessary) to the correct extent, you can place the resulting triangle exactly over any of the

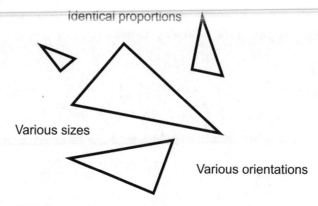

FIGURE 2-2 · Directly similar triangles.

other triangles. However, two triangles *are not* directly similar if we must flip one of them over, in addition to changing its size and rotating it, in order to place it exactly over the other one.

We call two triangles *inversely similar* if and only if they're directly similar when considered in the opposite rotational sense, or if they're directly similar after we flip one of them over. In other words, two triangles are inversely similar if and only if the "mirror image" of one is directly similar to the other.

Consider two directly similar triangles ΔABC and ΔDEF. We can symbolize the fact that they're directly similar by writing

$$\Delta ABC \sim \Delta DEF$$

The direct similarity symbol looks like a wavy minus sign. If the triangles ΔABC and ΔDEF are inversely similar, we have a more complicated situation, because it can arise in any of three different ways, as follows:

- Points D and E are transposed so ΔABC ~ ΔEDF.
- Points E and F are transposed so ΔABC ~ ΔDFE.
- Points D and F are transposed so ΔABC ~ ΔFED.

Congruent Triangles

Disagreement exists in mathematics literature concerning the meanings of the terms *congruence* and *congruent* for geometric figures in a plane. Some texts will tell you that two objects in a plane are congruent if and only if you can place one of them exactly over the other after a *rigid transformation* (rotating it or moving it around, but not flipping it over). Other texts define congruence to allow

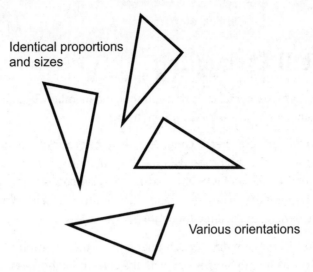

Identical proportions
and sizes

Various orientations

FIGURE 2-3 · Directly congruent triangles.

flipping-over, as well as rotation and motion. Let's stay away from that confusion
and formulate two separate definitions, one to account for either case.

Two triangles exhibit *direct congruence* (they're *directly congruent*) if and only
if they're directly similar, and the corresponding sides have identical lengths.
Figure 2-3 shows some examples. If you take one of the triangles and rotate it
clockwise or counterclockwise to the correct extent, you can "paste" it precisely
over any of the other triangles. Rotation and motion are allowed, but flipping-
over, also called *mirroring*, is forbidden.

Two triangles exhibit *inverse congruence* (they're *inversely congruent*) if and
only if they're inversely similar, and they also happen to be the same size.
Rotation and motion are allowed, and mirroring is *mandatory*.

If we have two triangles $\triangle ABC$ and $\triangle DEF$ that are directly congruent, we can
symbolize this fact by writing

$$\triangle ABC \cong \triangle DEF$$

The direct congruence symbol looks like an equals sign with a direct similarity
symbol on top. For two inversely congruent triangles $\triangle ABC$ and $\triangle DEF$, three
possibilities exist, as follows:

- Points D and E are transposed so $\triangle ABC \cong \triangle EDF$.
- Points E and F are transposed so $\triangle ABC \cong \triangle DFE$.
- Points D and F are transposed so $\triangle ABC \cong \triangle FED$.

Still Struggling

Here are four fundamental facts that you should remember about directly congruent triangles and inversely congruent triangles.

- If two triangles are directly congruent, then their corresponding sides have equal lengths as you proceed around both triangles in the same direction. The converse also holds true. If two triangles have corresponding sides with equal lengths as you proceed around them both in the same direction, then the two triangles are directly congruent.

- If two triangles are directly congruent, then their corresponding interior angles (the interior angles opposite the corresponding sides) have equal measures as you proceed around both triangles in the same direction. The converse does not necessarily hold true. Two triangles can have corresponding interior angles with equal measures when you proceed around them both in the same direction, and yet not be directly congruent.

- If two triangles are inversely congruent, then their corresponding sides have equal lengths as you proceed around the triangles in opposite directions. The converse also holds true. If two triangles have corresponding sides with equal lengths as you proceed around them in opposite directions, then the two triangles are inversely congruent.

- If two triangles are inversely congruent, then their corresponding interior angles have equal measures as you proceed around the triangles in opposite directions. The converse does not necessarily hold true. Two triangles can have corresponding interior angles with equal measures as you proceed around them in opposite directions, and yet not be inversely congruent.

Three-Point Principle

Let P, Q, and R represent three distinct and specific *noncollinear points* (meaning that we know exactly where they are, and they don't all lie on the same straight line, as shown in Fig. 2-4). The following statements hold true:

- P, Q, and R lie at the vertices of some triangle; let's call it W.
- W constitutes the only triangle having vertices P, Q, and R.

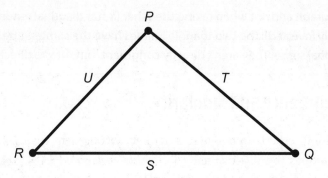

FIGURE 2-4 · The three-point principle and side-side-side triangles.

PROBLEM 2-1

Imagine a perfectly flat field completely enclosed by four straight lengths of fence. At their end points, the fences (which you can imagine as line segments as seen from high above the field) intersect at right angles. You build a straight fence diagonally across the field, dividing the field into two triangles. Are these triangles directly congruent? If not, are they directly similar?

SOLUTION

If you draw a diagram of this situation and examine it carefully, you'll see that the two triangles are directly congruent. Consider the theoretical images of the triangles (which, unlike the fences, you can move around in your imagination). You can rotate one of these theoretical triangles exactly 180° (π rad), either clockwise or counterclockwise, move it a little, and fit it exactly over the other one.

PROBLEM 2-2

Suppose that you have a telescope equipped with a camera. You focus on a distant, triangle-shaped road sign and take a photograph of it. Then you double the magnification of the telescope and, making sure the whole sign fits into the field of view of the camera, you take another photograph. When you look at the photographs on your computer screen, you see triangles in each photograph, of course. Are these triangles directly congruent? If not, are they directly similar?

SOLUTION

In the photos, one triangle looks larger than the other. But unless there's something wrong with the telescope, or you use a *star diagonal* when taking one photograph and not when taking the other (a star diagonal renders an image laterally inverted), the two triangle images have the same shape in the same rotational sense. They aren't directly congruent, but they're directly similar.

Criteria for Congruence and Similarity

We can use four criteria to define sets of directly congruent triangles. Geometers call these notions the *side-side-side* (SSS), *side-angle-side* (SAS), *angle-side-angle* (ASA), and *angle-angle-side* (AAS) principles. The last of these can also be called *side-angle-angle* (SAA). A fifth principle, called *angle-angle-angle* (AAA), can define sets of triangles that exhibit direct similarity, although they don't necessarily exhibit direct congruence.

Side-Side-Side (SSS)

Let S, T, and U represent defined, specific line segments. Let s, t, and u represent the lengths of S, T, and U, respectively. Suppose that S, T, and U meet at their end points P, Q, and R as shown in Fig. 2-4. In this situation, the following statements all hold true:

- Line segments S, T, and U determine a triangle W.
- W constitutes the only triangle with sides S, T, and U in this order, as you proceed around the triangle in the same rotational sense.
- All triangles having sides of lengths s, t, and u in this order, as you proceed around the triangles in the same rotational sense, are directly congruent.

Side-Angle-Side (SAS)

Let S and T represent defined, specific line segments. Let P represent a point that lies at the ends of both of these line segments. Denote the lengths of S and T by their lowercase counterparts s and t, respectively. Suppose that S and T form an angle x, expressed in the counterclockwise sense, at point P as shown in Fig. 2-5. In this case, the following statements all hold true:

- S, T, and x determine a triangle W.
- W constitutes the only triangle with sides S and T that form an angle x, measured counterclockwise from S to T, at point P.
- All triangles containing two sides of lengths s and t that form an angle x, measured counterclockwise from the side of length s to the side of length t, are directly congruent.

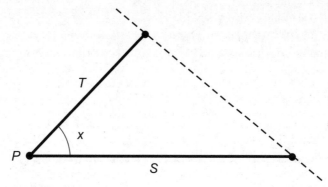

FIGURE 2-5 · Side-angle-side triangles.

Angle-Side-Angle (ASA)

Let *S* represent a line segment having length *s*, and whose end points are *P* and *Q*. Let *x* and *y* represent the angles formed relative to *S* by two lines *L* and *M* that run through *P* and *Q*, respectively (Fig. 2-6), such that we express both angles going counterclockwise. Then the following statements all hold true:

- *x*, *S*, and *y* determine a triangle *W*.
- *W* constitutes the only triangle determined by *x*, *S*, and *y*, proceeding from left to right.
- All triangles containing one side of length *s*, and whose other two sides form angles of *x* and *y* relative to the side whose length is *s*, with *x* on the left and *y* on the right and both angles expressed counterclockwise, are directly congruent.

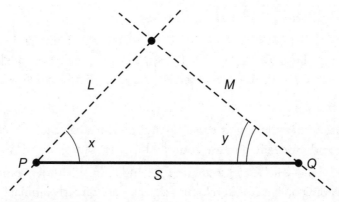

FIGURE 2-6 · Angle-side-angle triangles.

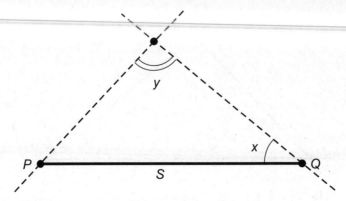

FIGURE 2-7 · Angle-angle-side triangles.

Angle-Angle-Side (AAS) or Side-Angle-Angle (SAA)

Let S represent a line segment having length s, and whose end points are P and Q. Let x and y represent angles, one adjacent to S and one opposite S, and both expressed in the counterclockwise sense (Fig. 2-7). The following statements all hold true:

- S, x, and y determine a triangle W.
- W constitutes the only triangle determined by S, x, and y in the counterclockwise sense.
- All triangles containing one side of length s, and two angles x and y, one adjacent and one opposite, expressed and proceeding in the counterclockwise sense, are directly congruent.

Angle-Angle-Angle (AAA)

Let L, M, and N represent lines that lie in a common plane and intersect in three points as illustrated in Fig. 2-8. Let x, y, and z represent the angles at these points, all expressed in the counterclockwise sense. The following statements all hold true:

- Infinitely many triangles exist having interior angles x, y, and z, in this order and proceeding in the counterclockwise sense.
- All triangles with interior angles x, y, and z, in this order, expressed and proceeding in the counterclockwise sense, are directly similar.

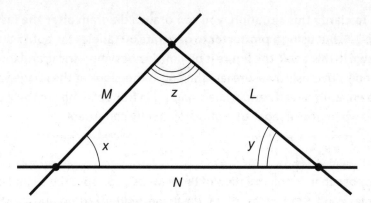

FIGURE 2-8 · Angle-angle-angle triangles.

Still Struggling

Do you wonder why we use the word "let" so often? For example, "Let P, Q, and R represent distinct points." Mathematicians write statements like this all the time. When an author suggests that you "let" things be a certain way, she asks you to *imagine* things that way, setting the scene for statements or problems that follow.

PROBLEM 2-3

Refer to Fig. 2-6 again. Suppose that the angles x and y both measure 60°. If we reverse the resulting triangle from left to right (we "flip it over" around a vertical axis), will the resulting triangle be directly similar to the original? Will it be directly congruent to the original?

SOLUTION

This problem illustrates a special case in which we can "flip over" a triangle and get another triangle that's not only inversely congruent, but also directly congruent, to the original. This coincidence occurs because the original triangle exhibits *bilateral symmetry*, meaning that it's symmetrical on either side of some defined straight-line axis.

To clarify this situation, you can draw a diagram after the fashion of Fig. 2-6, but using a protractor to generate 60° angles for both x and y. (As drawn in this book, the figure is obviously not symmetrical, and the angles x and y obviously measure less than 60°.) Then look at the image you have drawn, both directly and while standing in front of a mirror. You'll see that the two "mirror-image" triangles are directly congruent.

PROBLEM 2-4

Suppose that, in the situation of Problem 2-3, you split the triangle, whose angles x and y both measure 60°, down the middle by dropping a vertical line from the top vertex to the midpoint of line segment PQ. Are the resulting two triangles, each comprising half of the original, directly similar? Are they directly congruent? Are they inversely similar? Are they inversely congruent?

SOLUTION

These triangles constitute "mirror images" of each other, but you cannot magnify, reduce, and/or rotate one of them to make it fix exactly over the other one. The triangles are not directly similar, nor are they directly congruent, even though, in a sense, they have the same size and shape. They're inversely congruent, however, because they constitute equal-sized "mirror images" of each other. Because they're inversely congruent, we know that they're inversely similar.

TIP   *For two triangles to exhibit direct similarity, the lengths of their sides must exist in the same proportion, in order, as you proceed in the same rotational sense (counterclockwise or clockwise) around both triangles. In order to be directly congruent, their sides must have identical lengths, in order, as you proceed in the same rotational sense around both triangles.*

Types of Triangles

Let's categorize triangles broadly in a qualitative sense, that is, according to their qualities or characteristics.

Acute Triangle

We have an *acute triangle* if and only if each of the three interior angles is acute. In such a triangle, none of the angles measure as much as a right angle (90° or $\pi/2$ rad); they're all smaller than that. Figure 2-9 shows some examples.

All interior angles measure
less than 90°

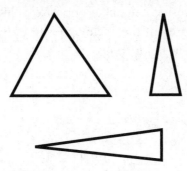

FIGURE 2-9 · In an acute triangle, all
angles measure less than a right
angle (90° or π/2 rad).

Obtuse Triangle

We have an *obtuse triangle* if and only if one of the three interior angles is obtuse, measuring more than a right angle (90° or π/2 rad) but less than a straight angle (180° or π rad). In a triangle of this type, the two nonobtuse angles are both acute. Figure 2-10 shows some examples.

One interior angle measures
more than 90°

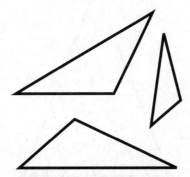

FIGURE 2-10 · In an obtuse triangle,
one angle measures more than a
right angle (90° or π/2 rad).

isosceles Triangle

Imagine a triangle with sides called S, T, and U that have lengths s, t, and u respectively. Let x represent the angle opposite S, let y represent the angle opposite T, and let z represent the angle opposite U. Suppose that *at least one* of the following equations holds true:

$$s = t$$
$$t = u$$
$$s = u$$
$$x = y$$
$$y = z$$
$$x = z$$

Figure 2-11 shows an example of such a situation, where $s = t$. Whenever we find a triangle that has two sides of identical length, we call it an *isosceles triangle*, and the following statements all hold true:

$$s = t \leftrightarrow x = y$$
$$t = u \leftrightarrow y = z$$
$$s = u \leftrightarrow x = z$$

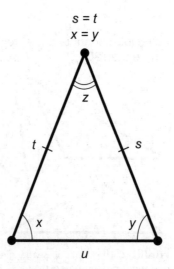

FIGURE 2-11 · An isosceles triangle.

Still Struggling

In a logical statement, a double-headed arrow (↔) stands for the expression "if and only if," and a single-headed arrow pointing to the right (→) stands for the expression "logically implies." For example, when we write

$$s = t \leftrightarrow x = y$$

we assert that

$$s = t \rightarrow x = y$$

and also that

$$x = y \rightarrow s = t$$

Mathematicians sometimes abbreviate "if and only if" as "iff," meaning that the logical implication works both ways. In mathematics and logic, when we claim that "A implies B," we mean "If A holds true, then B always holds true," or "If A, then B."

Equilateral Triangle

Imagine a triangle with sides called S, T, and U that have lengths s, t, and u respectively. Let x represent the angle opposite S, let y represent the angle opposite T, and let z represent the angle opposite U. Suppose that either of the following equations holds true:

$$s = t = u$$

or

$$x = y = z$$

In this case we have an *equilateral triangle* (Fig. 2-12), and we can make the logical statement

$$s = t = u \leftrightarrow x = y = z$$

Any two equilateral triangles chosen "at random" have precisely the same shape; they're directly similar. (As a matter of coincidence, they're inversely similar as well.)

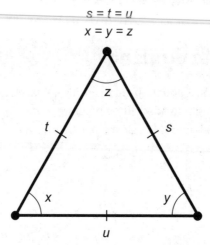

FIGURE 2-12 · An equilateral triangle.

Right Triangle

Imagine a triangle ΔPQR with sides S, T, and U, having lengths s, t, and u, respectively. This triangle constitutes a *right triangle* if and only if one of the interior angles is a right angle (90° or $\pi/2$ rad). Figure 2-13 illustrates a right triangle in which $\angle QRP$ forms the right angle. The side opposite the right angle has the longest length; we call it the *hypotenuse* of the right triangle. In Fig. 2-13, the hypotenuse has length u.

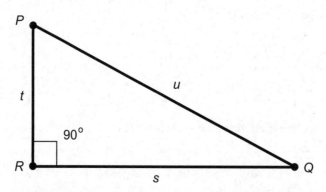

FIGURE 2-13 · A right triangle. The theorem of Pythagoras holds true for all such triangles.

Special Facts

Triangles have some special properties. These characteristics have applications in various branches of the physical sciences. You can expect to encounter these applications in science or engineering courses.

A Triangle Determines a Unique Plane

The vertex points of a specific triangle define one, and only one, Euclidean (flat) geometric plane. This fact should strike you as intuitively obvious when you give it a little thought. Try to imagine three points that don't all lie in the same plane! A specific Euclidean plane can contain infinitely many different triangles, but in such a case, all of the triangles, all of their sides, and all of their vertices are *coplanar* (meaning that they all lie in the same plane).

Sum of Angle Measures

In any triangle, the measures of the interior angles add up to a straight angle (180° or π rad). This fact holds true regardless of whether we have an acute, right, or obtuse triangle, as long as we express and measure all of the angles in the plane defined by the three vertices of the triangle.

Theorem of Pythagoras

Consider a right triangle defined by points P, Q, and R whose sides are S, T, and U having lengths s, t, and u, respectively. Let u represent the hypotenuse, as shown in Fig. 2-13. In this situation, the following equation, known as the *theorem of Pythagoras* or the *Pythagorean theorem* (named after the Greek philosopher who supposedly discovered it around the sixth century B.C.) always holds true:

$$s^2 + t^2 = u^2$$

TIP *Any triangle whose sides have lengths s, t, and u such that the foregoing equation holds true constitutes a right triangle.*

Perimeter of Triangle

Imagine a triangle defined by points P, Q, and R, and having sides S, T, and U of lengths s, t, and u, respectively, as shown in Fig. 2-14. We can calculate the perimeter B of the triangle with the formula

$$B = s + t + u$$

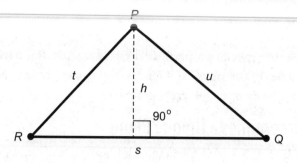

FIGURE 2-14 · Perimeter and area of a triangle.

Interior Area of Triangle

Consider the same triangle as defined above; refer again to Fig. 2-14. Let s represent the triangle's base length, and let h represent the triangle's height (the length of a perpendicular line segment between point P and side S). We can calculate the interior area A with the formula

$$A = sh/2$$

PROBLEM 2-5

Suppose that $\triangle PQR$ in Fig. 2-14 has sides of lengths $s = 10$ meters, $t = 7$ meters, and $u = 8$ meters. What's the perimeter B of this triangle?

✔ SOLUTION

We simply add up the lengths of the sides, obtaining

$$B = s + t + u$$
$$= (10 + 7 + 8) \text{ meters}$$
$$= 25 \text{ meters}$$

PROBLEM 2-6

Do any triangles exist having sides of lengths 10 meters, 7 meters, and 8 meters, in that order proceeding clockwise that *fail* to exhibit direct congruence with $\triangle PQR$ as described in Problem 2-5?

✔ SOLUTION

No. According to the side-side-side (SSS) principle, all triangles having sides of lengths 10 meters, 7 meters, and 8 meters, in this order as we proceed in the same rotational sense, are directly congruent.

QUIZ

Refer to the text in this chapter if necessary. A good score is eight correct. Answers are in the back of the book.

1. **Which of the following statements does not always hold true?**
   A. If two triangles have corresponding sides with equal lengths as we go around them both in the same direction, then the triangles are directly congruent.
   B. If two triangles are directly congruent, then their corresponding sides have equal lengths as we go around them both in the same direction.
   C. If two triangles are directly congruent, then their corresponding interior angles have equal measures as we go around them both in the same direction.
   D. If two triangles have corresponding interior angles with equal measures as we go around them both in the same direction, then the triangles are directly congruent.

2. **Upon casual observation, the triangle in Fig. 2-15 looks like**
   A. an acute triangle.
   B. an obtuse triangle.
   C. a reflex triangle.
   D. a right triangle.

3. **To find the area enclosed by the triangle shown in Fig. 2-15, we must multiply the length of line segment *RP* by**
   A. half of the length of line segment *RQ*.
   B. half of the length of line segment *QP*.
   C. half of the sum of the lengths of line segments *RQ* and *QP*.
   D. half of the shortest distance between point *Q* and line segment *RP*.

4. **All equilateral triangles are**
   A. directly congruent.
   B. inversely congruent.
   C. inversely similar.
   D. All of the above

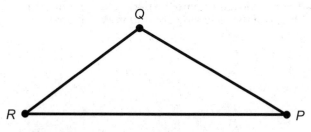

FIGURE 2-15 • Illustration for Quiz Questions 2 and 3.

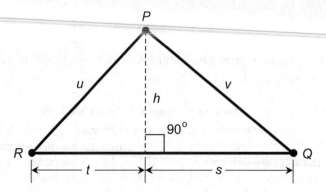

FIGURE 2-16 · Illustration for Quiz Questions 6 through 10.

5. We can use all of the following criteria to establish the direct congruence of two triangles, except for one. Which one?
   A. SAS
   B. SSS
   C. AAA
   D. ASA

6. In the situation of Fig. 2-16, we can have complete confidence that the triangle with sides measuring s, h, and v is
   A. an acute triangle.
   B. a right triangle.
   C. an equilateral triangle.
   D. an isosceles triangle.

7. We can mathematically determine the interior area A of the large triangle $\triangle QRP$ in Fig. 2-16 with one of the following equations. Which one?
   A. $A = hs$
   B. $A = ht$
   C. $A = (hu + hv)/2$
   D. $A = (hs + ht)/2$

8. We can mathematically determine the perimeter B of the large triangle $\triangle QRP$ in Fig. 2-16 with one of the following equations. Which one?
   A. $B = st + u + v$
   B. $B = s + t + u + v$
   C. $B = h(u + v)$
   D. $B = h + u + v$

9. Which of the following equations holds true for the situation of Fig. 2-16, based only on the information specifically given?

    A. $v^2 = s^2 + h^2$
    B. $(t+s)^2 = u^2 + v^2$
    C. $ts = u^2 + v^2$
    D. $h^2 = t^2 + u^2$

10. Which of the following equations holds true for the situation of Fig. 2-16, based only on the information specifically given?

    A. $uv = ts$
    B. $htu = hvs$
    C. $h^2 = u^2 - t^2$
    D. $t + s = u^2 - v^2$

chapter 3

Quadrilaterals

Within the confines of a Euclidean plane, we can call any four-sided geometric figure a *quadrilateral*. Inside a triangle, any given interior angle must measure more than 0° (0 rad) but less than 180° (π rad); with a quadrilateral, any given interior angle must measure more than 0° (0 rad) but less than 360° (2π rad).

CHAPTER OBJECTIVES

In this chapter, you will

- Define and name quadrilaterals.
- Classify quadrilaterals according to general shape.
- Learn the relationships between sides and angles of quadrilaterals.
- Break quadrilaterals into triangles.
- Calculate quadrilateral perimeters.
- Calculate the areas enclosed by quadrilaterals.

Types of Quadrilaterals

We can categorize any four-sided plane figure as a *square*, a *rhombus*, a *rectangle*, a *parallelogram*, a *trapezoid*, or a *general quadrilateral*. Let's define these terms and the look at some examples.

Requirements

A four-sided geometric object *must* have four properties to "qualify" as a Euclidean plane quadrilateral:

- All four vertices must lie in the same plane.
- All four sides must constitute straight line segments of finite length.
- No side can have zero length or negative length.
- No two sides can intersect except at their end points.

We can't let a quadrilateral "stray" out of a single plane. We can't allow sides to have any curvature whatsoever. We can't "stretch" a side to infinite length or "crush" it down to a point having no length at all. A true plane quadrilateral cannot have any side whose length we define as negative.

The vertices of a triangle must inevitably lie in a single geometric plane, because any three points, no matter which ones we choose, define a unique geometric plane. But when we choose four points "at random" in space, they don't all necessarily lie in the same plane.

TIP   *Any three points in space lie in a single plane, but a fourth one can get "out of alignment." That's why a four-legged stool or table often wobbles, and why it's so difficult to trim the lengths of the legs so the wobbling stops. Once the ends of the legs lie in a single plane so that they define the vertices of a plane quadrilateral, the stool or table won't wobble as long as the floor remains perfectly flat. (If the floor isn't flat, you have a real prescription for frustration with four-legged stools, but a three-legged stool will stand firm even on irregular terrain.)*

Square

A square has four sides, all measuring the same length. In addition, all of the interior angles measure 90° ($\pi/2$ rad). Figure 3-1 shows the general situation. The length of each side equals s units. There exists no limit as to how large s can become, but it must always have a positive, nonzero value.

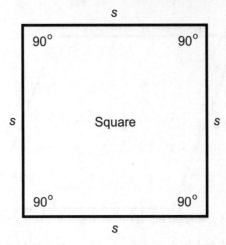

All four sides have equal length;
all four angles are right angles

FIGURE 3-1 · Example of a square. Sides have length *s*, and the interior angles all constitute right angles (90° or π/2 rad).

Rhombus

In a rhombus, all four sides have the same length, but the angles don't all have to measure 90°. A "generic" rhombus looks something like the polygon shown in Fig. 3-2. All four sides have length *s*. Opposite pairs of angles have equal measure, but adjacent pairs of angles can (and usually do) differ. In this illustration, the two angles labeled *x* have equal measure, as do the two angles labeled *y*. In a rhombus, both pairs of opposite sides are parallel.

TIP   *A square constitutes a special type of rhombus in which all four angles happen to have the same measure.*

Rectangle

In a rectangle, all four angles have equal measure, but the sides don't necessarily all have equal lengths. A "generic" rectangle looks something like the polygon shown in Fig. 3-3. All four angles measure 90° (π/2 rad). Opposite pairs of sides are equally long, but adjacent pairs of sides usually differ in length. In the case of Fig. 3-3, the two sides labeled *s* have equal lengths, as do the two sides labeled *t*.

TIP   *A square constitutes a special type of rectangle in which all four sides happen to have equal lengths.*

All four sides have equal length;
opposite angles have equal measure

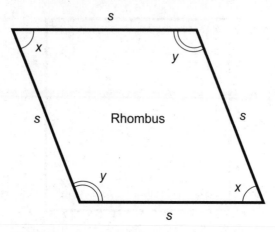

FIGURE 3-2 • Example of a rhombus. Sides have length *s*, while *x* and *y* denote interior angle measures.

Opposite sides have equal length;
all four angles are right angles

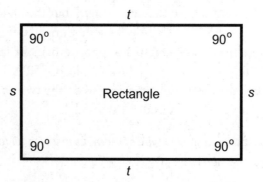

FIGURE 3-3 • Example of a rectangle. Sides have lengths *s* and *t*, while the interior angles all constitute right angles (90° or π/2 rad).

Parallelogram

We can define a parallelogram according to its outstanding characteristic: Both pairs of opposite sides are parallel. That's it! That quality alone allows a plane quadrilateral to qualify as a parallelogram. Whenever both pairs of opposite sides in a Euclidean plane quadrilateral are parallel, those pairs also have the

Opposite sides have equal length;
opposite angles have equal measure

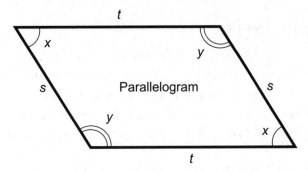

FIGURE 3-4 · Example of a parallelogram. Sides have lengths *s* and *t*, while *x* and *y* denote interior angle measures.

same length. In addition, pairs of opposite angles have equal measure. Figure 3-4 shows an example of a parallelogram in which both angles labeled *x* have equal measure, both angles labeled *y* have equal measure, both sides labeled *s* have the same length, and both sides labeled *t* have the same length.

TIP   *A rectangle constitutes a special sort of parallelogram. So does a rhombus, and so does a square.*

Trapezoid

We can define a trapezoid as a plane quadrilateral in which one pair of opposite sides is parallel. Otherwise, no restrictions exist (other than the ones necessary to ensure that we have a "legitimate" Euclidean plane quadrilateral). Figure 3-5

Two opposite sides are parallel;
no other constraints exist

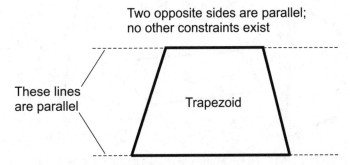

FIGURE 3-5 · In a trapezoid, one pair of opposite sides is parallel.

shows an example of a trapezoid. The dashed lines represent the parallel lines in which the two parallel sides of the trapezoid lie.

General Quadrilaterals

In a general quadrilateral, we don't have to impose any restrictions on the lengths of the sides, although no interior angle's measure can stray outside of the range 0° (0 rad) to 360° (2π rad), noninclusive. As long as all four vertices lie in the same plane, no two sides intersect except at their end points, and all four sides of the figure are straight line segments of finite and positive length, we're okay.

Irregular Quadrilaterals

We can consider any quadrilateral "general." A rectangle, for example, is a specific type of general quadrilateral. So is a rhombus; so is a trapezoid. But we can find plenty of general quadrilaterals that don't fall into any of the foregoing categories. They don't exhibit any symmetry or apparent orderliness. We call four-sided polygons of the "maverick type" *irregular quadrilaterals*. Figure 3-6 shows some examples.

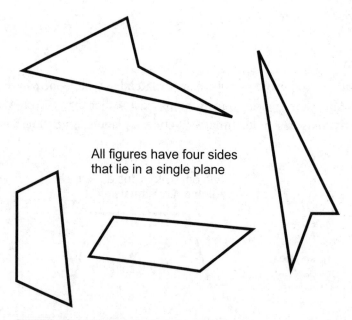

All figures have four sides
that lie in a single plane

FIGURE 3-6 · Examples of irregular quadrilaterals. The sides can all
have different lengths, and the angles can all have different measures.

PROBLEM 3-1

What type of quadrilateral constitutes the boundaries (end lines and side-lines) of a football field?

SOLUTION

Assuming the groundskeepers do their job correctly, a football field has the shape of a rectangle. All four corners form right angles (90°). In addition, both pairs of opposite sides are equally long. That is to say, the two sidelines have the same length, as do the two end lines.

PROBLEM 3-2

Suppose that we define a quadrilateral *ABCD* so that we encounter the vertex points *D*, *C*, *B*, and *A* in that order going clockwise around the figure. Suppose further that we have

$$\angle CBA = \angle ADC$$

and

$$\angle BAD = \angle DCB$$

What specific things can we say about this quadrilateral?

SOLUTION

I recommend that you draw a diagram to illustrate this situation, because most people can't directly envision these constraints "in their mind's eyes." You'll see that $\angle CBA$ lies opposite $\angle ADC$, and $\angle BAD$ lies opposite $\angle DCB$. The fact that opposite pairs of angles have equal measure tells you that the quadrilateral constitutes a parallelogram. It might be a special type of parallelogram such as a rhombus, rectangle, or square, but you can have absolute confidence that it's a parallelogram no matter what.

Facts about Quadrilaterals

Every quadrilateral has certain properties, depending on the "species." Following are some useful facts concerning four-sided Euclidean plane figures.

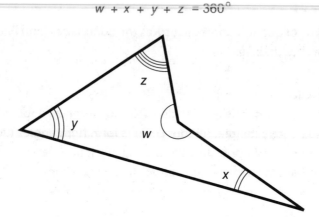

$$w + x + y + z = 360°$$

FIGURE 3-7 · In any plane quadrilateral, the sum of the measures of the interior angles w, x, y, and z equals a full circle (360° or 2π rad).

Sum of Measures of Interior Angles

As long as all four sides of a quadrilateral are straight line segments of positive and finite length and as long as all four vertices lie in the same plane, the measures of the four interior angles always add up to 360° (2π rad). Figure 3-7 shows an example of an irregular quadrilateral. We denote the interior angles as w, x, y, and z. In this particular example, angle w measures more than 180° (π rad). The other three angles are all acute. If you enlarge Fig. 3-7 and use a protractor to measure the interior angles and if you then add up all four angle measures, you should obtain 360° (within the margin of observation error).

Still Struggling

You might call the irregular quadrilateral in Fig. 3-7 a "boomerang" or a "distorted arrowhead," although neither of these is an "official" geometry term. You might also call this object a "reflex quadrilateral" because angle w is a reflex angle or a "concave quadrilateral" because the figure has a "dent." Feel free to make up your own terms in geometry once in awhile, but use caution! Your readers might not know what you mean unless you explain it to them.

Parallelogram Diagonals

Consider a parallelogram defined by four vertex points P, Q, R, and S, which we encounter in that order as we proceed clockwise around the figure. Let D represent a line segment connecting the "nearer pair" of vertices P and R as shown in Fig. 3-8A. In this situation, the line segment D constitutes the *minor diagonal* of the parallelogram, and

$$\Delta PQR \cong \Delta RSP$$

Let E represent a line segment connecting the "farther pair" of vertices Q and S as shown in Fig. 3-8B. In this case, the line segment E constitutes the *major diagonal* of the parallelogram, and

$$\Delta QRS \cong \Delta SPQ$$

Remember that the equals sign with the wavy line above it translates to the phrase "is directly congruent to"!

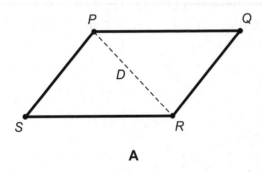

A

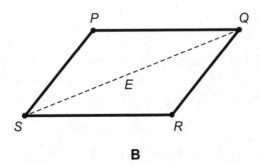

B

FIGURE 3-8 • Triangles defined by the minor diagonal (A) or the major diagonal (B) of a parallelogram are congruent.

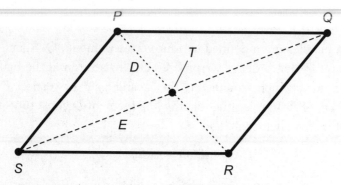

FIGURE 3-9 · The diagonals of a parallelogram bisect each other.

Bisection of Parallelogram Diagonals

Suppose that we have a parallelogram defined by four vertex points P, Q, R, and S, which we encounter in that order as we proceed clockwise around the figure. Let D represent the minor diagonal connecting P and R; let E represent the major diagonal connecting Q and S (Fig. 3-9). In this scenario, the two line segments D and E bisect each other at their intersection point T. In addition, we have

$$\Delta PQT \cong \Delta RST$$

and

$$\Delta QRT \cong \Delta SPT$$

The converse of the foregoing statement also holds true: If we have a plane quadrilateral whose diagonals bisect each other, then that quadrilateral constitutes a parallelogram.

Rectangle

Consider a parallelogram defined by four points P, Q, R, and S, which we encounter in that order as we go clockwise around the figure. Suppose that any of the following statements holds true for angles in degrees:

$$\angle SRQ = 90° = \pi/2 \text{ rad}$$

$$\angle PSR = 90° = \pi/2 \text{ rad}$$

$$\angle QPS = 90° = \pi/2 \text{ rad}$$

$$\angle RQP = 90° = \pi/2 \text{ rad}$$

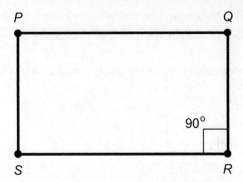

FIGURE 3-10 · If a parallelogram has one right interior angle, then the parallelogram constitutes a rectangle.

In this situation, all four interior angles are right angles, and the parallelogram is therefore a rectangle (a four-sided plane polygon whose interior angles all have equal measures). The converse of this statement also holds true: If a quadrilateral is a rectangle, then any given interior angle is a right angle. Figure 3-10 shows an example of a parallelogram $PQRS$ in which $\angle SRQ = 90° = \pi/2$ rad. Because one angle is a right angle and opposite pairs of sides are parallel, all four of the angles must measure 90° ($\pi/2$ rad).

Rectangle Diagonals

Imagine a parallelogram defined by four points P, Q, R, and S, which we encounter in that order as we go clockwise around the figure. Let D represent the diagonal connecting P and R; let E represent the diagonal connecting Q and S. Suppose that D and E are equally long, as shown in Fig. 3-11. In that case,

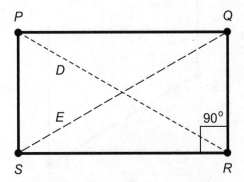

FIGURE 3-11 · The diagonals of a rectangle have equal length.

the parallelogram is a rectangle. The converse of this statement also holds true: If a parallelogram is a rectangle, then its two diagonals are equally long.

TIP *A parallelogram constitutes a rectangle* **if and only if** *its diagonals have equal lengths.*

Rhombus Diagonals

Suppose we have a parallelogram defined by four points P, Q, R, and S, which we encounter in that order as we go clockwise around the figure. Let D represent the diagonal connecting P and R; let E represent the diagonal connecting Q and S. If D runs perpendicular to E as shown in Fig. 3-12, then the parallelogram is a rhombus. The converse of this statement also holds true: If a parallelogram is a rhombus, then its major and minor diagonals run perpendicular to each other.

TIP *A parallelogram constitutes a rhombus* **if and only if** *its diagonals are perpendicular.*

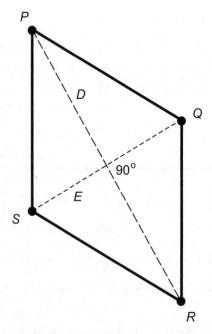

FIGURE 3-12 ∙ The diagonals of a rhombus intersect at right angles.

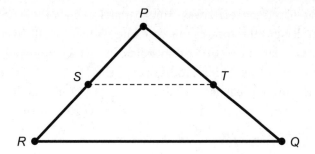

FIGURE 3-13 • A trapezoid is formed by "chopping off" the top of a triangle.

Trapezoid within a Triangle

Consider a triangle defined by three points P, Q, and R, which we encounter in that order as we go clockwise around the figure. Let S represent the midpoint of side PR, and let T represent the midpoint of side PQ as shown in Fig. 3-13. In this case, line segments ST and RQ run parallel to each other, and the figure $STQR$ defined by the four vertex points S, T, Q, and R (in order going clockwise) constitutes a trapezoid. In addition, the length of line segment ST equals half the length of line segment RQ.

Median of a Trapezoid

Consider a trapezoid defined by four points P, Q, R, and S, which we encounter in that order as we go clockwise around the figure. Let T represent the midpoint of side PS, and let U represent the midpoint of side QR as shown in Fig. 3-14. We call line segment TU the *median* of trapezoid $PQRS$.

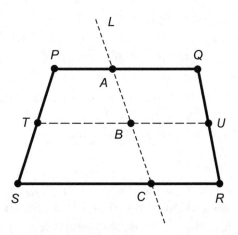

FIGURE 3-14 • The median of a trapezoid, also showing a transversal line.

The median of a trapezoid always runs parallel to both the base (in this case line segment *SR*) and the top (in this case line segment *PQ*). Also, the median splits the trapezoid into two other trapezoids. In this scenario, polygons *PQUT* and *TURS* are both trapezoids. Additionally, the length of line segment *TU* equals half the sum of the lengths of line segments *PQ* and *SR*. In more general terms, the length of a trapezoid's median is the average of the lengths of the base and the top.

Still Struggling

Recall from your pre-algebra course that you can find the average, also called the *arithmetic mean*, of two numbers by adding them and then dividing the result by 2.

Median with Transversal

Look again at Fig. 3-14. Suppose that *L* represents a transversal line that crosses both the top of the large trapezoid (line segment *PQ*) and its base (line segment *SR*). The transversal line *L* also crosses the large trapezoid's median, line segment *TU*. Let *A* represent the point at which *L* crosses *PQ*, let *B* represent the point at which *L* crosses *TU*, and let *C* represent the point at which *L* crosses *SR*. In this case, line segments *AB* and *BC* have equal length.

Still referring to Fig. 3-14, suppose that *PQRS* is a trapezoid, with sides *PQ* and *RS* parallel. Let *TU* represent a line segment parallel to both *PQ* and *RS*, and that intersects both of the nonparallel sides of the trapezoid (sides *PS* and *QR*). Let *L* represent a transversal line that crosses all three parallel line segments *PQ*, *TU*, and *RS*, at the points *A*, *B*, and *C*, respectively, as shown. In this scenario, line segment *TU* is the median of the large trapezoid *PQRS* if and only if line segments *AB* and *BC* have equal length.

PROBLEM 3-3

Suppose that a four-sided plane figure has diagonals that both measure the same length, and, in addition, they intersect at a right angle. What can we say about this polygon?

✔ SOLUTION

Based on the rules that we've learned so far, the figure must constitute a rectangle, because its diagonals have equal lengths. It must also be a rhombus, because its diagonals run perpendicular to each other. Only one type of polygon—a square—can exist as a rectangle and a rhombus "at the same time."

TIP *A square is a special type of rhombus in which both pairs of opposite interior angles have the same measure. A square is also a special type of rectangle in which both pairs of opposite sides have equal lengths.*

PROBLEM 3-4

Suppose that a sign manufacturing company gets tired of making rectangular billboards and decides to put up a trapezoidal billboard instead. The top and the bottom of the billboard run horizontally, but neither of the other sides runs vertically. The big sign measures 20 meters across the top edge and 30 meters across the bottom edge. Two different corporations want to advertise on the billboard, and their chief executives both insist on having portions of equal height. What's the length of the line that divides the spaces allotted to the two advertisements? Does this compromise represent a "fair" or "equitable" division of the sign?

✔ SOLUTION

The line segment that divides the two portions constitutes the median of the sign. Its length, therefore, equals 25 meters, which is the average of 20 meters and 30 meters. We can debate whether or not this particular division represents a "fair" or "equitable" apportionment of the sign. The advertiser on the bottom receives more sign surface area than the advertiser on the top gets, but casual passersby might more easily notice the advertisement on the top.

Perimeters and Interior Areas

The *perimeter* of a polygon equals the sum of the lengths of all its sides. We can also define a polygon's perimeter as the distance going exactly once around the whole edge of the figure, starting at some point on one of its sides and proceeding

~~clockwise or counterclockwise until we reach that point again.~~ The *interior area* of a plane polygon quantifies of the size of the region enclosed by the figure in the same plane as its vertices and sides.

TIP   *We must always express perimeter values in **linear units** (or, if you prefer, "plain old units"), and interior-area values in **square units** (or "units squared").*

Perimeter of Parallelogram

Consider a parallelogram defined by points P, Q, R, and S, which we encounter in that order as we go clockwise around the figure. Suppose that the opposite pairs of sides have lengths d and e as shown in Fig. 3-15. The two angles labeled x have equal measure. Let d represent the base length, and let h represent the height. We can calculate the perimeter B of the parallelogram with the formula

$$B = 2d + 2e$$

Interior Area of Parallelogram

Suppose that we have a parallelogram as defined earlier and in Fig. 3-15. The interior area A equals the product of the base length and the height. We can calculate it using the simple formula

$$A = dh$$

Perimeter of Rhombus

Imagine a rhombus defined by points P, Q, R, and S, which we encounter in that order as we go clockwise around the figure. The rhombus constitutes a

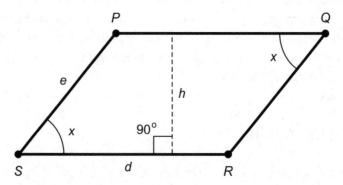

FIGURE 3-15 · Perimeter and area of a parallelogram. The parallelogram constitutes a rhombus if and only if $d = e$.

special case of the parallelogram (Fig. 3-15) in which all four sides are equal (so, in the case of Fig. 3-15, we have $d = e$). Let d equal the length of any one side. We can calculate the perimeter B of the rhombus using the formula

$$B = 4d$$

Interior Area of Rhombus

Consider a rhombus as defined earlier and in Fig. 3-15, where $d = e$. Let's denote the length of any one side as d. The interior area A of the rhombus equals the product of the side length and the height. We can calculate it with the formula

$$A = dh$$

Perimeter of Rectangle

Consider a rectangle defined by points P, Q, R, and S, which we encounter in that order as we go clockwise around the figure. Imagine that the sides measure d and e as shown in Fig. 3-16. Let d represent the base length, and let e represent the height. We can calculate the rectangle's perimeter B with the formula

$$B = 2d + 2e$$

Interior Area of Rectangle

Consider a rectangle as defined earlier and in Fig. 3-16. We can calculate the rectangle's interior area A with the formula

$$A = de$$

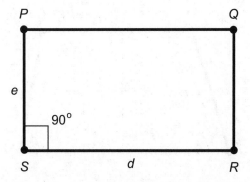

FIGURE 3-16 • Perimeter and area of a rectangle. The figure constitutes a square if and only if $d = e$.

Perimeter of Square

Imagine a square defined by points P, Q, R, and S, and having sides all of the same length. The square constitutes a special case of the rectangle (Fig. 3-16) in which $d = e$. Let's denote the lengths of all four sides as d. We can calculate the square's perimeter B with the formula

$$B = 4d$$

Interior Area of Square

Consider a square as defined earlier and in Fig. 3-16, where $d = e$. Let's denote the lengths of all four sides as d. We can calculate the square's interior area A by squaring the length of any side. We have the formula

$$A = d^2$$

Perimeter of Trapezoid

Imagine a trapezoid defined by points P, Q, R, and S, which we encounter in that order as we go clockwise around the figure. Imagine that the sides have lengths d, e, f, and g as shown in Fig. 3-17. Let d represent the base length, let h represent the height, let x represent the angle between the sides having lengths d and e, and let y represent the angle between the sides having lengths d and g. Suppose that the sides having lengths d and f (line segments RS and PQ) are parallel. We can calculate the trapezoid's perimeter B with the formula

$$B = d + e + f + g$$

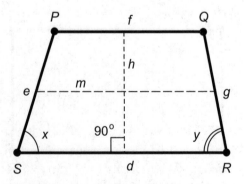

FIGURE 3-17 · Perimeter and area of a trapezoid based on its various dimensions.

Interior Area of Trapezoid

Consider a trapezoid as defined earlier and in Fig. 3-17. The interior area A equals the average of the lengths of the base and the top, multiplied by the height. We can calculate A using the formula

$$A = [(d + f)/2]\, h$$
$$= (dh + fh)/2$$

Now suppose that m represents the length of the median of the trapezoid, that is, a line segment parallel to the base and the top, and midway between them. The interior area A equals the product of the length of the median and the height. We can use the formula

$$A = mh$$

PROBLEM 3-5

Refer back to Problem 3-4. Suppose that the whole billboard measures 15 meters high. It's a trapezoidal billboard, measuring 20 meters along the top edge and 30 meters along the bottom. We divide the sign by placing a horizontal median midway between the top and the bottom. What fraction of the total billboard surface area, as a percentage, does the advertiser with the top half get?

SOLUTION

The length of the median, as determined in Problem 3-4, equals 25 meters, the average of the lengths of the bottom and the top. Therefore $m = 25$. We're told that $h = 15$. We can calculate the total interior area of the sign—call it A_{total}—as follows:

$$A_{total} = 25 \text{ meters} \times 15 \text{ meters}$$
$$= 375 \text{ meters squared}$$

We calculate the area of the top half by considering the trapezoid for which m constitutes the base. Let's use the more complicated formula—the one involving the arithmetic mean, above—in order to find the interior area of this smaller trapezoid. We can call its area A_{top}. The base length of this trapezoid equals 25 meters, while the top measures 20 meters long. The height

~~equals 7.5 meters, half the height of the whole sign. We calculate A_{top} as~~
follows:

$$A_{top} = [(25 \text{ meters} + 20 \text{ meters})/2] \times 7.5 \text{ meters}$$
$$= (45 \text{ meters}/2) \times 7.5 \text{ meters}$$
$$= 22.5 \text{ meters} \times 7.5 \text{ meters}$$
$$= 168.75 \text{ meters squared}$$

The fraction of the total area represented by the top portion of the sign equals the ratio of A_{top} to A_{total}. That's 168.75 meters squared divided by 375 meters squared, or 0.45. The top advertiser gets 45/100, or 45%, of the total sign area.

PROBLEM 3-6

Suppose that the billboard constitutes a rectangle rather than a trapezoid, measuring 25 meters across both the top and the bottom. Suppose the sign is 15 meters tall, and we want to split it into upper and lower portions, one for each of two different advertisers, Top Inc. and Bottom Inc. Suppose that the executives of Bottom Inc. demand that the Top Inc. only get 45% of the total area of the sign because of Top Inc.'s more favorable viewing position. How far from the bottom of the sign should we place the dividing line?

SOLUTION

The total area of the sign, A_{total}, equals the product of the base (or top) length and the height, as follows:

$$A_{total} = 25 \text{ meters} \times 15 \text{ meters}$$
$$= 375 \text{ meters squared}$$

This figure equals the total area that we found in Solution 3-5. Therefore, 45% of this, A_{top}, is the same as it was then: 168.75 meters squared. We can now calculate the area of the bottom portion, A_{bottom}, as follows:

$$A_{bottom} = A_{total} - A_{top}$$
$$= (375 - 168.75) \text{ meters squared}$$
$$= 206.25 \text{ meters squared}$$

Let x represent the distance, in meters, of the dividing line from the sign's bottom edge. In that case, x represents the lengths of the two vertical sides of the bottom rectangle. We already know that the dividing line (which constitutes the top edge of the bottom rectangle) measures 25 meters long, as does the base. According to all this information, we can use the following formula to define the area of the bottom portion:

$$A_{bottom} = 25x$$

We know that A_{bottom} = 206.25 meters squared. We can plug this value into the above equation to get

$$206.25 = 25x$$

When we divide each side of this equation by 25 meters, we obtain

$$x = (206.25 \text{ meters squared})/(25 \text{ meters})$$
$$= 8.25 \text{ meters}$$

We should place the dividing line 8.25 meters above the bottom edge of the billboard.

Still Struggling

Does the previously-determined placement represent a fair division of the sign's total area? The lawyers for Top Inc. and Bottom Inc. could decide the matter in court, doubtless at shareholder expense. As mathematicians, we'd better stay out of the dispute!

QUIZ

Refer to the text in this chapter if necessary. A good score is eight correct. Answers are in the back of the book.

1. Figure 3-18 illustrates a trapezoid *PQRS* whose top edge measures 7 units wide, bottom edge measures 9 units wide, and height equals 6 units. Based on this information, what's the perimeter of trapezoid *PQRS*?

   A. 22 units
   B. 24 units
   C. 48 units
   D. We need more information to answer this question.

2. Based on the information shown in Fig. 3-18, what's the interior area of trapezoid *PQRS*?

   A. 54 units squared
   B. 48 units squared
   C. 42 units squared
   D. We need more information to answer this question.

3. Based on the information shown in Fig. 3-18, what's the interior area of triangle *SQR*?

   A. 27 units squared
   B. 24 units squared
   C. 21 units squared
   D. We need more information to answer this question.

4. Based on the information shown in Fig. 3-18, what's the interior area of triangle *PQS*?

   A. 27 units squared
   B. 24 units squared

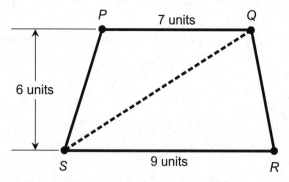

FIGURE 3-18 · Illustration for Quiz Questions 1 through 4.

C. 21 units squared

D. We need more information to answer this question.

5. **A square is a special type of**

A. rhombus.

B. parallelogram.

C. rectangle.

D. All of the above

6. **Suppose that in the situation of Fig. 3-19, angle *x* measures π/2 rad. In that case, we can be certain that**

A. all four sides of polygon *PQRS* are equally long.

B. angle *SPQ* measures 45°.

C. angle *RSP* measures 135°.

D. All of the above

7. **We can find the interior area of a rectangle by**

A. multiplying the lengths of any two adjacent sides.

B. multiplying the lengths of any two opposite sides.

C. adding up the lengths of all four sides.

D. multiplying the lengths of all four sides and then dividing by 2.

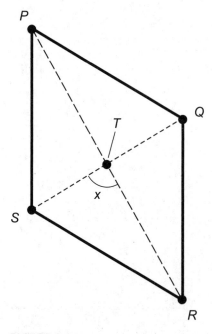

FIGURE 3-19 · Illustration for Quiz Question 6.

8. When we encounter a trapezoid, we can have complete confidence that the sum of the measures of the interior angles

    A. exceeds 2π rad.
    B. is less than 2π rad.
    C. equals 2π rad.
    D. depends on the relative lengths of the edges.

9. When we encounter a parallelogram, we can have complete confidence that the measures of either pair of opposite interior angles

    A. adds up to π rad.
    B. adds up to 2π rad.
    C. adds up to $\pi/2$ rad.
    D. None of the above

10. In a plane quadrilateral, the measure x of any particular interior angle must lie within a certain range. How can we express that range?

    A. 0 rad $< x < \pi/2$ rad
    B. 0 rad $< x < 2\pi$ rad
    C. 0 rad $< x < \pi$ rad
    D. 0 rad $< x < 3\pi/2$ rad

chapter 4

Other Plane Figures

There exists no limit to the number of straight sides (also called edges) and vertices (points where the sides join at their ends) that a plane polygon can possess. More complicated objects can have curved sides or edges. Let's explore the properties of general Euclidean plane figures.

CHAPTER OBJECTIVES

In this chapter, you will

- Define and classify diverse plane figures.
- Learn the relationships among the sides and angles of regular polygons.
- Calculate the perimeters and interior areas of regular polygons.
- Evaluate the characteristics of circles and ellipses.
- Observe a regular polygon as the number of sides grows without limit.

Five Sides and Up

For a geometric object to "qualify" as a Euclidean plane polygon, it must have several characteristics, as follows:

- All of its vertices must lie in the same plane.
- No two sides may cross over each other.
- No two vertices may coincide.
- No three vertices may lie on the same straight line.
- The sides must all constitute straight line segments having finite, positive length.

The Regular Pentagon

Figure 4-1 shows a five-sided plane polygon, all of whose sides have the same length, and all of whose interior angles have the same measure. We call this figure a *regular pentagon*. It constitutes a *convex* figure because its exterior never "bends inward." In a convex plane polygon, every interior angle has a measure of less than 180° (π rad).

The Regular Hexagon

A convex plane polygon with six sides, all of which have equal length, is called a *regular hexagon* (Fig. 4-2). If we take a large number of equal-sized regular hexagons, we can place them neatly together without any gaps. (Have you ever visited an old-fashioned barbershop where the floor comprised thousands of hexagonal tiles that fit snugly up against each other?) This property makes the regular hexagon

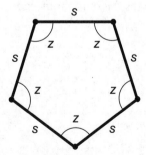

FIGURE 4-1 • A regular pentagon. Each side has length *s*, and each interior angle has measure *z*.

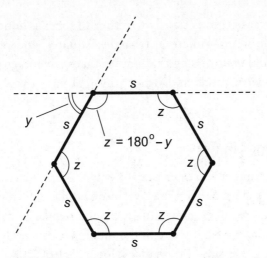

FIGURE 4-2 · A regular hexagon. Each side has length s, and each interior angle has measure z. The extensions of sides (dashed lines) are the subject of Problem 4-1.

a special sort of figure, along with the equilateral triangle, the square, and the regular octagon. Certain crystalline solids form regular hexagonal shapes when they fracture. Snowflakes, for example, have components with this shape.

The Regular Octagon

Figure 4-3 shows a *regular octagon*. It's a convex plane polygon with eight sides, all equally long, and eight interior angles, all of equal measure. We can fit large

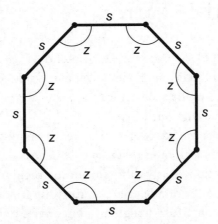

FIGURE 4-3 · A regular octagon. Each side has length s, and each interior angle has measure z.

numbers of regular octagons tightly together to form a "honeycomb matrix," just as we can do with equilateral triangles, squares, and regular hexagons. We should not find it surprising that nature takes advantage of this property, building octagonal crystals and other physical structures in the material universe.

Regular Polygons in General

For every whole number n greater than or equal to 3, we can construct a regular polygon with n sides. So far we've seen the equilateral triangle ($n = 3$), the square ($n = 4$), the regular pentagon ($n = 5$), the regular hexagon ($n = 6$), and the regular octagon ($n = 8$). If we're in the mood, we can easily imagine a regular polygon with 1,000 sides (a "regular kilogon"), 1,000,000 sides (a "regular megagon"), or 1,000,000,000 sides (a "regular gigagon"). In a regular plane polygon, no matter how many sides it has, the measure of any given individual interior angle must always be less than 180° (π rad).

TIP   *As the number of sides in a regular polygon increases without limit, the measures of the individual interior angles approach 180°(πrad), and the figure approaches a circle. In fact, a "regular gigagon" (as defined above) would look like a perfect circle, even if we examined it under a microscope! All the sides, vertices, and angles would seem to "merge" into a continuous, symmetrical, convex curve.*

General, Many-Sided Polygons

Once we remove the restrictions concerning the relationship among the sides of a polygon having four sides or more, the potential for variety increases without limit. In a general Euclidean plane polygon, the sides can have all different lengths, and the measure of each interior angle can range anywhere from 0° (0 rad) to 360° (2π rad), noninclusive.

Figure 4-4 shows some examples of general, many-sided polygons. The object at the top left is a *regular nonconvex octagon* whose sides all have equal length, but that obviously differs from the regular octagon we usually imagine. Four of the interior angles are acute; four are reflex. The other two objects in Fig. 4-4 constitute irregular, nonconvex polygons. All three of these objects nevertheless share the essential characteristics of a Euclidean plane polygon.

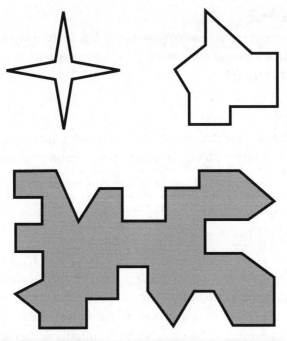

FIGURE 4-4 • General, many-sided polygons. The object with the shaded interior is the subject of Problem 4-2.

PROBLEM 4-1

What's the measure of each interior angle of a regular hexagon?

✔ SOLUTION

Draw a horizontal line segment to start. All the other sides must form exact duplicates of this initial side, but rotated with respect to the first line segment by whole-number multiples of a certain angle. The rotation angle from side to side equals 360° divided by 6 (a full rotation divided by the number of sides), or 60°. Imagine the lines on which two adjacent sides lie. Look back at Fig. 4-2. These lines subtend a 60° angle with respect to each other, if you look at the acute angle y between the dashed lines. But if you look at the obtuse angle z, you'll find that it measures 120°, which equals 180° − y. This obtuse angle z constitutes an interior angle of the hexagon. Therefore, each interior angle of a regular hexagon measures 120°.

PROBLEM 4-2

Briefly glance at the lowermost polygon in Fig. 4-4 (the one with the shaded interior). Don't look at it for more than 2 seconds. How many sides do you suppose that it has?

SOLUTION

Most people underestimate the number of sides in complicated plane figures like this. After you've made your guess, count the sides. How far off were you?

Some Rules for Polygons

All plane polygons share certain characteristics. We can calculate the perimeter or area of any polygon, no matter how complicated (although we might appreciate a computer's power to help us solve a particularly messy problem of this sort). Specific rules and definitions apply to the interior and exterior angles, and also to the relationships among the angles and the sides.

It's Greek to Us

Mathematicians, scientists, and engineers often use Greek letters to represent geometric angles. The most common symbol for this purpose is an italicized, lowercase Greek letter theta (θ), as we learned in Chap. 2.

When we want to write about two different angles, we can use a second Greek letter along with θ. Mathematicians often choose the italicized, lowercase letter phi (pronounced "fie" or "fee"), which looks like a lowercase English letter "o" leaning to the right, with a forward slash through it (ϕ). You should get used to seeing these symbols. If you have much to do with mathematics, engineering, or science in the future, you're going to encounter them a lot.

Sometimes the italic, lowercase Greek alpha ("AL-fuh"), beta (BAY-tuh"), and gamma ("GAM-uh") are used to represent angles. These letters, respectively, look like this: α, β, γ. When things get messy and we have many angles to talk about, we might use numeric subscripts with a single Greek letter. As you carry on in mathematics and science, you'll occasionally see angles denoted in a form such as θ_1, θ_2, θ_3, and so on.

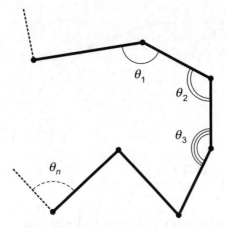

FIGURE 4-5 • Adding up the measures of
the interior angles of a general, many-sided
polygon.

Sum of Interior Angles

Consider a plane polygon having n sides. Let θ_1, θ_2, θ_3, ..., θ_n represent the
interior angles, as shown in Fig. 4-5. The following equation holds for angular
measures expressed in degrees:

$$\theta_1 + \theta_2 + \theta_3 + \cdots + \theta_n = 180n - 360$$

$$= 180(n - 2)$$

If we express the angular measures in radians, then

$$\theta_1 + \theta_2 + \theta_3 + \cdots + \theta_n = \pi n - 2\pi$$

$$= \pi(n - 2)$$

Still Struggling

In the previous two examples, we've left out the degree symbol (°) and the
radian abbreviation (rad) for simplicity. We can get away with this shortcut as
long as we ensure that we (and our readers) know which angular units apply in
any given situation.

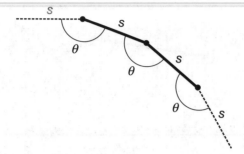

FIGURE 4-6 · Interior angles of a regular, many-sided polygon.

Individual Interior Angles of Regular Polygon

Consider a plane polygon having n sides, whose interior angles all have equal measure given by θ, and whose sides all have equal length given by s (Fig. 4-6). This figure constitutes a regular Euclidean plane polygon, and we can calculate the measure of each interior angle θ in degrees with the formula

$$\theta = (180n - 360)/n$$

If we express the angular measures in radians, then

$$\theta = (\pi n - 2\pi)/n$$

Exterior Angles

We can express or measure an *exterior angle* of a polygon going counterclockwise from a specific side to the extension of the side immediately to the left. Figure 4-7 shows an example of this process. If the arc of the angle lies outside

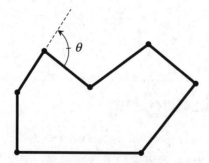

FIGURE 4-7 · Exterior angle of an irregular polygon. We express the angle θ going counterclockwise from a given side to the line containing the adjacent side on the left.

the polygon, then the resulting angle θ has a measure between, but not including, 0° and 180°. The angle has positive measure because we express it while rotating in the "positively counterclockwise" sense. Symbolically, we can say that

$$0° < \theta < 180°$$

Perimeter of Regular Polygon

Consider a regular plane polygon having n sides of length s, with vertices P_1, P_2, P_3, ..., P_n as shown in Fig. 4-8. We can calculate the perimeter B of the polygon using the formula

$$B = ns$$

TIP *Some of the following rules involve* **trigonometry.** *Six* **trigonometric func-** **tions,** *also known as* **circular functions,** *exist: the* **sine (sin), cosine (cos), tangent** **(tan), cosecant (csc), secant (sec),** *and* **cotangent (cot).** *All six of these functions produce specific "output" numbers when you "feed" them specific angular "input" numbers. You can find the sine of an angle by entering the angle's measure in degrees or radians into a calculator and then hitting the "sine" or "sin" function key. You can find the cosine by entering the angle's measure in degrees or radians and then hitting "cosine" or "cos." Some calculators have a "tangent" or "tan" function key, and others don't. If your calculator doesn't have a tangent key, you can find the tangent of an angle by dividing its sine by its cosine. Many calculators*

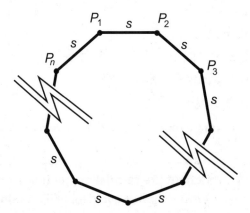

FIGURE 4-8. • Perimeter and area of a regular, n-sided polygon. Points P_1, P_2, P_3, ..., P_n constitute the vertices. Each side has length s.

lack a "cotangent" or "cot" key, but the cotangent of an angle equals the recipro-
cal of its tangent, or the cosine divided by the sine. The cosecant equals the recip-
rocal of the sine. The secant equals the reciprocal of the cosine.

Interior Area of Regular Polygon

Consider a regular, n-sided polygon, each of whose sides have length s as defined earlier and in Fig. 4-8. If we express the interior angles in degrees, then we can calculate the polygon's interior area A with the formula

$$A = (ns^2/4) \cot (180/n)$$

If we express the interior angles in radians, then

$$A = (ns^2/4) \cot (\pi/n)$$

PROBLEM 4-3

What's the interior area of a regular, 10-sided polygon, each of whose sides measures exactly 2 units long? Express the answer to two decimal places.

SOLUTION

In this case, $n = 10$ and $s = 2$. Let's use degrees for the angles, so that we can plug our values of n and s into the first formula, above, and proceed as follows:

$$A = (10 \times 2^2/4) \cot (180/10)$$
$$= (10 \times 4/4) \cot 18$$
$$= 10 \cot 18$$
$$= 10 \cos 18 / \sin 18$$
$$= 10 \times 0.951057/0.309017$$
$$= 10 \times 3.07769$$
$$= 30.78 \text{ square units (rounded off to two decimal places)}$$

In order to obtain an answer to two decimal places, we can use five or six decimal places throughout the calculation, rounding off only at the end. This precaution will ensure that we avoid (or at least minimize) *cumulative rounding errors.*

PROBLEM 4-4

What's the interior area of a regular, 100-sided polygon, each of whose sides measures exactly 0.20 units long? Express the answer to two decimal places.

SOLUTION

In this example, $n = 100$ and $s = 0.20$. If you're astute, you'll notice that the perimeter of this polygon equals $100 \times 0.20 = 20$ units, the same as the perimeter of the 10-sided polygon of Problem 4-3, which is $10 \times 2.0 = 20$ units. Imagine these two regular polygons sitting side by side. Draw approximations of them if you like. It seems reasonable to suppose that the area of the 100-sided polygon should slightly exceed that of the 10-sided figure. Let's find out, using radians instead of degrees this time. Let $\pi = 3.14159$.

Set your calculator to work with radians, not degrees, before each and every use of a trigonometric function key. Here we go:

$$A = (100 \times 0.20^2/4) \cot (\pi/100)$$

$$= (100 \times 0.04/4) \cot 0.0314159$$

$$= \cot 0.0314159$$

$$= \cos 0.0314159 / \sin 0.0314159$$

$$= 0.999507/0.031411$$

$$= 31.82 \text{ square units (rounded off to two decimal places)}$$

TIP   *Whenever you execute calculations such as the two foregoing, you should go through the entire process twice or more. It's amazing how many errors people make when using calculators to do plain arithmetic. The most common mistakes occur as a result of pressing one or more function keys in the wrong order. However, once in awhile something else happens; a speck of dirt might get into one of the calculator keys, for example, causing that key to "think" you've hit it two or three times when in fact you've hit it only once!*

Circles and Ellipses

We can define a *circle* as a geometric figure consisting of all points in a plane that lie *equidistant* (i.e., equally far away) from a specified center point. Imagine a flashlight with a round lens that produces a brilliant central beam surrounded by a dim cone of light. Suppose that you switch this flashlight on and then point it straight down at the floor in a dark room. The outline of the dim light cone constitutes a circle. If you turn the flashlight so that the entire dim light cone lands on the floor but the brilliant central light ray does not point straight down, the outline of the dim light cone forms an *ellipse*. All circles and ellipses represent examples of *conic sections*. This term arises from the fact that we can define both the circle and the ellipse as sets of points resulting from the intersection of a flat, two-dimensional plane with a three-dimensional cone.

A Special Number

The perimeter (more often called the circumference) of a circle, divided by its diameter in the same units, equals a constant independent of the size of the circle, as long as we remain in a single geometric plane when we make our measurements. Mathematicians first noticed this fact thousands of years ago. They spent centuries trying to determine the exact value of this constant, settling for some time on the approximate value of 22/7. Today, we know that we can't express this constant precisely as a ratio of whole numbers. For this reason, we call it an *irrational number*. (In this context, the term "irrational" means "having no ratio.") If we try to write this constant as a decimal expression, we get a nonterminating, nonrepeating sequence of digits after the decimal point. We call the constant *pi*, and symbolize it using the lowercase Greek letter having that name (π). It's the same constant π that we encountered when we defined the radian as a unit of angular measure.

TIP *Supercomputers have calculated the value of π to millions of decimal places. It equals approximately 3.14159. If you need more accuracy, you can use the calculator function in a personal computer. Set the program for radians, not degrees, and then find the Arccosine (also known as the inverse cosine and sometimes symbolized cos⁻¹) of the integer −1. You'll get a display of π to all the digits your computer's calculator program can handle. In theory, π rad equals precisely the Arccosine of −1. You might want to memorize this fact, so that you can always "bring up" an accurate value of π on your calculator (unless, of course, it has a "pi" key built in!).*

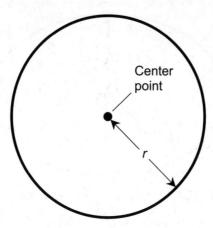

FIGURE 4-9 · Dimensions of a circle. The radius measures r units.

Circumference of Circle

Consider a circle having radius r as shown in Fig. 4-9. We can calculate the circle's circumference B with the formula

$$B = 2\pi r$$

Interior Area of Circle

Once again, consider the circle defined earlier and illustrated in Fig. 4-9. We can calculate the interior area A of the circle with the formula

$$A = \pi r^2$$

Approximate Circumference of Ellipse

Imagine an ellipse whose "long radius" (technically called the *major semiaxis* or the *semimajor axis*) measures r_1 units and whose "short radius" (called the *minor semiaxis* or the *semiminor axis*) measures r_2 units as shown in Fig. 4-10. We can approximate the circumference B of this figure using the formula

$$B = 2\pi \left[(r_1^2 + r_2^2)/2\right]^{1/2}$$

where the 1/2 power of a quantity represents the positive square root of that quantity. The above formula provides the best accuracy when the lengths of the semiaxes don't differ by much. As the semiaxis lengths grow more different from each other, the formula gets less precise.

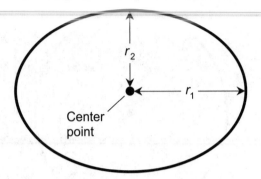

FIGURE 4-10 • Dimensions of an ellipse. The major semiaxis measures r_1 units, and the minor semiaxis measures r_2 units.

TIP *We need calculus to determine the circumference of an ellipse exactly. That mathematical discipline lies beyond the scope of this course.*

Interior Area of Ellipse

Once again, imagine an ellipse whose major semiaxis measures r_1 units and minor semiaxis measures r_2 units (Fig. 4-10). We can calculate the ellipse's interior area A with the formula

$$A = \pi r_1 r_2$$

Ellipticity

The ratio of the length of an ellipse's major semiaxis to the length of its minor semiaxis tells us how much the ellipse is elongated, or "out of round." We call the ratio r_1/r_2 the *ellipticity*, often symbolized by the lowercase, italic Greek letter epsilon (ε). Symbolically, we have

$$\varepsilon = r_1/r_2$$

Still Struggling

When $\varepsilon = 1$, we have a special case where an ellipse constitutes a perfect circle. Because we define r_1 as the major (longer) semiaxis, ε is always greater than or equal to 1. Don't confuse ellipticity with *eccentricity*, an entirely different measure of the extent to which a curve deviates from a perfect circle.

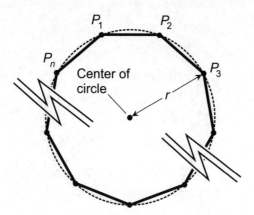

FIGURE 4-11 · Perimeter and area of inscribed regular polygon. The radius of the circle measures r units. Vertices of the polygon, all of which lie on the circle, are $P_1, P_2, P_3, ..., P_n$.

Perimeter of Inscribed Regular Polygon

Consider a regular plane polygon having n sides, and whose vertices $P_1, P_2, P_3, ..., P_n$ lie on a circle of radius r (Fig. 4-11). If we specify angles in degrees, then we can calculate the perimeter B of the polygon with the formula

$$B = 2nr \sin (180/n)$$

If we express angles in radians, then

$$B = 2nr \sin (\pi/n)$$

Interior Area of Inscribed Regular Polygon

Consider a regular polygon as defined earlier and in Fig. 4-11. If we express angles in degrees, then we can calculate the interior area A of the polygon as

$$A = (nr^2/2) \sin (360/n)$$

If we express the angles in radians, then

$$A = (nr^2/2) \sin (2\pi/n)$$

Perimeter of Circumscribed Regular Polygon

Imagine a regular plane polygon having n sides whose center points $P_1, P_2, P_3, ..., P_n$ lie on a circle of radius r (Fig. 4-12). If we express angles in degrees, then we can calculate the perimeter B of the polygon as

$$B = 2nr \tan (180/n)$$

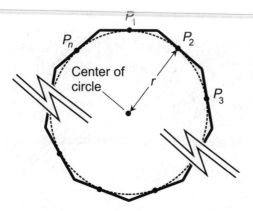

FIGURE 4-12 · Perimeter and area of circumscribed regular polygon. The radius of the circle measures *r* units. Center points of the sides of the polygon, all of which lie on the circle, are P_1, P_2, P_3, ..., P_n.

If angles are given in radians, then

$$B = 2nr \tan (\pi/n)$$

Interior Area of Circumscribed Regular Polygon

Consider a regular polygon as defined earlier and in Fig. 4-12. If we specify angles in degrees, then we can calculate the interior area *A* of the polygon with the formula

$$A = nr^2 \tan (180/n)$$

If angles are specified in radians, then

$$A = nr^2 \tan (\pi/n)$$

Perimeter of Circular Sector

Imagine a certain *sector* of a circle of radius *r*, shown by the heavy outlined "pizza-pie slice" in Fig. 4-13. Let θ represent the apex angle, as shown, in radians. We can calculate the perimeter *B* of the sector in linear units using the formula

$$B = 2r + r\theta$$

If we specify θ in degrees, then the perimeter *B* of the sector in linear units is

$$B = 2r (1 + 90\theta)/\pi$$

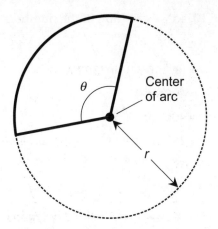

FIGURE 4-13 · Perimeter and area of circular sector. The radius of the circle measures *r* units, and the arc subtends an angle *θ*.

Interior Area of Circular Sector

Once again, imagine a sector of a circle as defined earlier and in Fig. 4-13. Let *θ* represent the apex angle in radians. We can calculate the sector's interior area *A* in square units with the formula

$$A = r^2\theta/2$$

If we specify *θ* in degrees, then the interior area *A* of the sector in square units is

$$A = 90\ r^2\theta/\pi$$

PROBLEM 4-5

What's the area of a regular octagon inscribed within a circle whose radius equals precisely 10 units?

SOLUTION

Let's use the formula for the area of an inscribed regular polygon, where angles are expressed in degrees:

$$A = (nr^2/2)\ \sin\ (360/n)$$

In this formula, *A* represents the area in square units, *n* represents the number of sides in the polygon, and *r* represents the radius of the circle. We know that *n* = 8 (because we have a regular octagon, or eight-sided polygon) and $r = 10$, so we can plug in the numbers and use a calculator to obtain

$$A = (8 \times 10^2/2) \sin (360/8)$$
$$= 400 \sin 45$$
$$= 400 \times 0.7071$$
$$= 283 \text{ square units (approximately)}$$

PROBLEM 4-6

What's the perimeter of a regular 12-sided polygon circumscribed around a circle whose radius is exactly 4 units?

SOLUTION

Let's use the formula for the perimeter of a circumscribed regular polygon, where angles are expressed in radians:

$$B = 2nr \tan (\pi/n)$$

Here, B represents the perimeter, n represents the number of sides in the polygon, and r represents the radius of the circle. Consider $\pi = 3.14159$. We know that $n = 12$ and $r = 4$. We plug in the numbers and use a calculator, making sure to set the angle function for radians (not degrees). We get

$$B = 2 \times 12 \times 4 \tan (\pi/12)$$
$$= 96 \tan 0.261799$$
$$= 96 \times 0.26795$$
$$= 25.72 \text{ units (approximately)}$$

PROBLEM 4-7

How should we expect the perimeter of the circumscribed polygon in Problem 4-6 to compare with the circumference of the circle around which it's circumscribed?

SOLUTION

We can reasonably imagine that the perimeter of the polygon slightly exceeds the circle's circumference. Let's calculate the circumference to test this hunch using the formula for the circumference of a circle, as follows:

$$B = 2\pi r$$

where B represents the circumference and r represents the radius. We know that $r = 4$, and we consider $\pi = 3.14159$. Therefore

$$B = 2 \times 3.14159 \times 4$$
$$= 25.13 \text{ units (approximately)}$$

That's a little less than the perimeter of the circumscribed polygon, just as we thought.

TIP   *Suppose that we* circumscribe *a circle with a regular polygon* P_c *having* n *sides (where* n *represents a positive integer larger than 3), and then we increase* n *without limit. Also suppose that we* inscribe *the same circle with another regular polygon* P_i *having the same number of sides as* P_c *at all times. As* n *grows larger indefinitely,* P_c *and* P_i *become more and more nearly the same and they both approach the circle in terms of perimeter and interior area. The measures of the interior angles approach 180° (π rad). The lengths of the sides approach zero.*

QUIZ

Refer to the text in this chapter if necessary. A good score is eight correct. Answers are in the back of the book.

1. Consider a *regular septagon* (a regular plane polygon having seven sides of equal length and seven interior angles of equal measure). What's the measure of each individual interior angle?
   A. $3\pi/4$ rad
   B. $5\pi/7$ rad
   C. $7\pi/8$ rad
   D. $13\pi/14$ rad

2. An individual interior angle in a *regular* plane polygon always measures less than
   A. $\pi/2$ rad.
   B. π rad.
   C. $\pi/4$ rad.
   D. $3\pi/4$ rad.

3. An individual angle in *any* plane polygon always measures less than
   A. $\pi/2$ rad.
   B. π rad.
   C. $3\pi/2$ rad.
   D. 2π rad.

4. Consider a regular plane polygon having n sides of equal length and interior angles all of equal measure. As we've learned, we can calculate the measure θ of each individual interior angle with the formula

   $$\theta = (180n - 360)/n$$

   Based on this information, we can see that as we increase the number of sides in a regular plane polygon indefinitely, the *sum of the measures* of all the interior angles
   A. approaches 360°.
   B. approaches $(360^2)°$ or 129,600°.
   C. increases without limit.
   D. approaches zero.

5. Which of the following characteristics tells us that a given figure *does not* constitute a plane polygon?
   A. No three vertices lie along a single line.
   B. All of the vertices lie in a single plane.
   C. All of the sides constitute line segments of finite, nonzero length.
   D. Two of the sides cross over each other.

6. Figure 4-14 illustrates an ellipse. Suppose that the indicated dimensions are exact. What's the area of the ellipse, rounded off to three decimal places? Consider $\pi = 3.14159$.

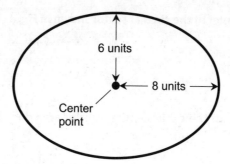

FIGURE 4-14 · Illustration for Quiz Questions 6 and 7.

    A. 43.982 square units
    B. 150.796 square units
    C. 153.938 square units
    D. We need calculus to figure it out.

7. **What's the approximate circumference of the ellipse of Fig. 4-14, rounded off to one decimal place?**
    A. 37.7 units
    B. 41.4 units
    C. 50.3 units
    D. 44.4 units

8. **Figure 4-15 illustrates a circular sector (heavy solid lines and curve). Suppose that the indicated dimensions are exact. What's the interior area of the sector?**
    A. 12 square units
    B. 4 square units
    C. 16 square units
    D. 20 square units

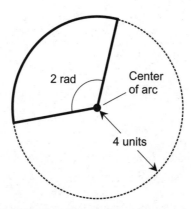

FIGURE 4-15 · Illustration for Quiz Questions 8 through 10.

9. What's the perimeter of the circular sector shown in Fig. 4-15?
   A. 12 units
   B. 14 units
   C. 16 units
   D. 20 units

10. What proportion of the circle's entire interior area does the sector shown in Fig. 4-15 represent?
    A. $1/\pi$
    B. $\pi/10$
    C. $3/(7\pi)$
    D. $2/(3\pi)$

chapter **5**

Compass and Straight Edge

In geometry, the term *construction* refers to a drawing that we can make using simple tools, with the intent of demonstrating a certain principle. Constructions can serve as a powerful learning technique, because they force you to think about the properties of geometric objects, independent of numerical lengths and angle measures. Constructions can also provide some challenging games!

CHAPTER OBJECTIVES

In this chapter, you will

- Draw generic circles, lines, rays, and line segments.
- Construct angles and arcs.
- Bisect line segments and angles.
- Construct perpendicular and parallel line segments.
- Duplicate line segments and angles.

Tools and Rules

The most common type of geometric construction requires two instruments, both of which you can purchase at any office supply store. One instrument lets you draw circles, and the other lets you draw straight line segments. Once you have these tools, you can use them only according to certain "rules of the game."

Drafting Compass

The *drafting compass* allows you to draw circles of various sizes based on specific center points. It has two straight shafts joined by a hinge. One shaft ends in a sharp point that can't mark anything, but that you can "stick" into a piece of paper to serve as an "anchor." The other shaft has brackets in which you mount a pencil. When you want to draw a circle, you press the sharp point down on a piece of paper (with some cardboard underneath to protect the table or desk top), open the hinge to get the desired radius, bring the pencil to the paper, and draw the circle by rotating the whole assembly at least once around. You can draw arcs by rotating the compass partway around.

TIP   *For geometric constructions, the compass must not have an angle measurement scale at its hinge. If it has a scale that indicates angle measures or otherwise quantifies the extent of its spread, you must ignore that scale.*

Straight Edge

A *straight edge* helps you to draw line segments by placing a pencil against the object and running it alongside. A conventional ruler will work for this purpose, but it's not the best tool for formal geometric constructions because it has a calibrated scale. You're better off using a *drafting triangle*. Use any edge of the triangle as the straight edge. You can even use a stiff piece of cardboard with a known straight side, such as the back of a writing tablet after you've used up the paper.

TIP   *Ignore the angles at the apexes of a drafting triangle. Some drafting triangles have two 45° angles and one 90° angle; others have one 30° angle, one 60° angle, and one 90° angle. You mustn't take advantage of these standard angle measures when performing geometric constructions, so it doesn't matter which type of drafting triangle you use.*

What's Allowed

With a compass, you can draw circles or arcs having any radius you want (up to the maximum that the device will create, of course). You can choose the center point "at random," or you can place the sharp tip of the compass down on a predetermined, existing point and define that point as the center of the circle or arc.

You can adjust a compass to replicate the distance between any two defined points by setting the nonmarking tip down on one point and the end of the pencil down on the other point, and then holding the compass setting constant.

With the straight edge, you can draw line segments of any length, up to the entire length of the tool. You can draw a "random" line segment, choose a specific point through which the line segment passes, or connect any two specific points with a line segment.

What's Not Allowed

Whatever sort of circle or line segment you draw, you must never try to measure the radius or the length against a calibrated scale of any kind. You may not measure angles using a calibrated device. You may not make any reference marks on either the compass or the straight edge. Marking on a straight edge constitutes "cheating," but referencing a distance using a compass is okay, even though the two acts might seem qualitatively identical.

Still Struggling

As you do a geometric construction, you might wonder if you can "legally" imagine the result of infinitely many operations or infinitely many repetitions of a single operation. The answer is no, you may not do that! You mustn't repeat a maneuver, or any set of maneuvers, "forever" to geometrically approach a desired result, and then claim that result as a valid construction. You must physically complete the whole operation in a finite number of steps.

Creating Points

To define an arbitrary point, you can simply draw a dot on the paper anywhere you want. Alternatively, you can set the nonmarking point of the compass down on the paper, in preparation for drawing an arc or circle centered at an arbitrary point. You can also define points wherever two line segments intersect, wherever an arc or circle intersects a line segment, or wherever an arc or circle intersects another arc or circle.

Drawing Line Segments

You can construct line segments "at random," through any point, starting at any point, through any two points, or connecting any two points.

When you want to draw an arbitrary line segment, place the straight edge down on the paper and run a pencil along the edge as shown in Fig. 5-1A. You can make the line segment as long or as short as you want but never longer than the length of the straight edge. If you want to draw a line segment longer than the straight edge, don't align the straight edge with part of the line segment and then try to extend it. Use a longer straight edge, so that you can create the entire segment in one "swipe."

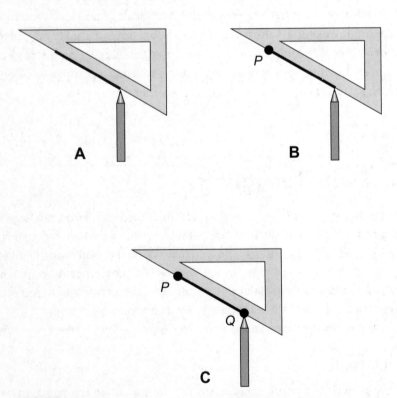

FIGURE 5-1 · At A, construction of an arbitrary line segment. At B, construction of a line segment starting at a single predetermined point. At C, construction of a line segment connecting two predetermined points.

When you want to draw a line segment through, or starting at, a single defined point, place the tip of the pencil on that point (call it point *P*), place the straight edge down against the point of the pencil, and then run the pencil back and forth along the edge. If you want the point to constitute an end point of the line segment, run the pencil away from the point in one direction as shown in Fig. 5-1B.

When you want to draw a line segment through two defined points (call them *P* and *Q*), place the tip of the pencil on one of the points, place the straight edge down against the tip of the pencil, rotate the straight edge until it lines up with the other point while still firmly resting against the tip of the pencil, and then run the pencil back and forth along the edge so that the mark passes through both points. If you want the points to lie at the ends of the line segment, make sure that the pencil makes its mark only between the points and not past them on either side (Fig. 5-1C).

Denoting Rays

To denote a ray, you must first locate or choose the end point of the ray (call it point *P*). Then place the tip of the pencil at the end point and place the straight edge against the tip of the pencil. Orient the straight edge in the direction you want the ray to go. Move the tip of the pencil away from the point in the direction of the ray, as far as you want without running off the end of the straight edge (Fig. 5-2A). Finally, draw an arrow at the end of the line segment opposite the starting point *P* (Fig. 5-2B). The arrow indicates that you want the ray to extend infinitely in that direction.

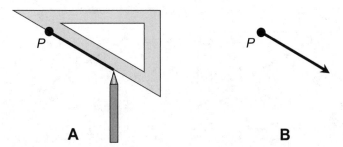

FIGURE 5-2 · Construction of a ray. First, construct a line segment ending at a point (A); then put an arrow at the end opposite the point (B).

Denoting Lines

In order to draw a line, follow the same procedure as you would to draw a line segment. Then place arrows at both ends (Fig. 5-3). You can construct a line "at random" (as shown in Fig. 5-3A and B), through a single defined point P (as shown in Fig. 5-3C and D) or through two defined points P and Q (as shown in Fig. 5-3E and F).

Drawing Circles

To draw a circle around a "random" point, place the nonmarking tip of the compass down on the paper, set the compass to the desired radius, and rotate

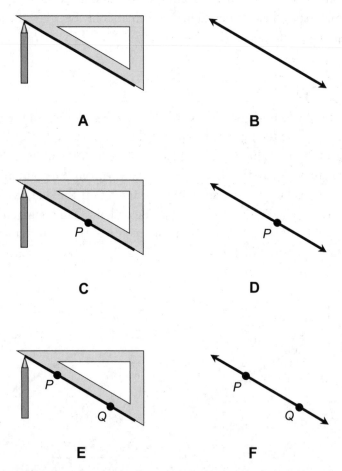

FIGURE 5-3 · At A and B, construction of an arbitrary line. At C and D, construction of a line through a single predetermined point. At E and F, construction of a line through two predetermined points.

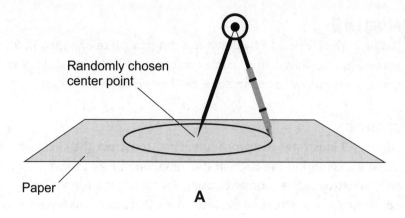

Randomly chosen
center point

Paper

A

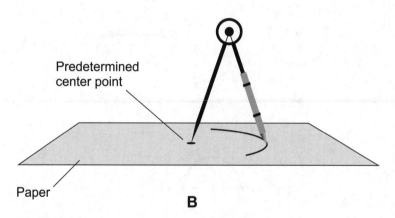

Predetermined
center point

Paper

B

FIGURE 5-4 • At A, construction of a circle centered on an arbitrary point. At B, construction of an arc centered at a predetermined point.

the instrument through a full circle (Fig. 5-4A). If you have a predetermined center point (marked by a dot), place the nonmarking tip down on the dot and rotate the instrument through a full circle.

Drawing Arcs

To draw an arc centered at a random point, place the nonmarking tip of the compass down on the paper, set the compass to the desired radius, and rotate the instrument through the desired arc. If you have a predetermined center point (marked by a dot), place the nonmarking tip down on the dot and rotate the instrument through the desired arc (Fig. 5-4B).

Define a specific point *P* by drawing a dot on a piece of paper. Then, with your compass, draw a small circle centered at *P*. Now construct a second circle, concentric with the first one, but having twice the radius.

SOLUTION

Figure 5-5 illustrates the procedure. First, construct the circle with your compass, centering the circle at the initial point *P* as shown in Fig. 5-5A. Then construct a line segment *L* using your straight edge, with one end at point *P* and passing through the circle at another point, which you can call *Q*.

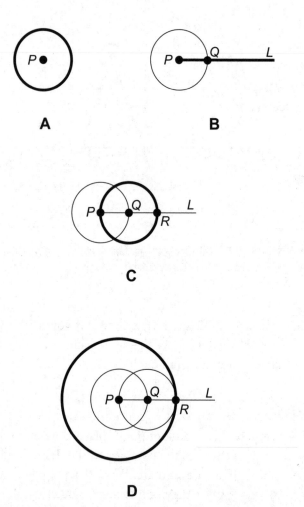

FIGURE 5-5 • Illustration for Problem 5-1.

Extend *L* outside the circle for a distance considerably greater than the circle's radius (Fig. 5-5B). Next, construct a second circle, centering it at point *Q* and leaving the compass set for the same radius as it was when you drew the original circle. This new circle intersects *L* at point *P* (the center of the original circle) and also at a new point *R* (Fig. 5-5C). Next, place the nonmarking tip of the compass back at point *P* and open up the compass so that the pencil tip lands on point *R*. Finally, draw a new circle centered at point *P*, with a radius equal to the length of line segment *PR* (as shown in Fig. 5-5D).

PROBLEM 5-2

Draw three points on a piece of paper, placed in such a way that they don't all lie along the same line. Label the points *P*, *Q*, and *R*. Construct Δ*PQR* connecting these three points. Draw a circle whose radius equals the length of side *PQ*, but that's centered at point *R*.

SOLUTION

Figure 5-6 shows the process. First, put down and label the initial points as shown in Fig. 5-6A. Then connect the points with line segments to construct Δ*PQR* (Fig. 5-6B). Next, place the nonmarking tip of your compass at point *Q* and the tip of the pencil on point *P*. (If you want, you can construct a small arc through *P* as shown in Fig. 5-6C, demonstrating that you've got the compass opened up to the correct span.) With the compass thereby set to define the length of line segment *PQ*, place the nonmarking tip of the compass on point *R*. Finally, as shown in Fig. 5-6D, construct the full circle centered at point *R*.

PROBLEM 5-3

Can we "legally" place the nonmarking tip of the compass at point *P* and then place the pencil tip to draw an arc through point *Q*, in order to define the length of line segment *PQ* in Problem 5-2?

SOLUTION

Yes, we can. This method works just as well as the procedure defined in the solution to Problem 5-2.

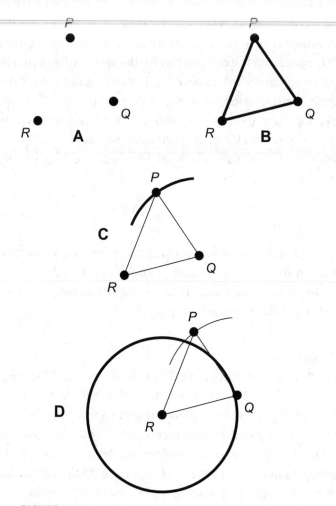

FIGURE 5-6 · Illustration for Problems 5-2 and 5-3.

Linear Construction Methods

The following paragraphs describe how to perform various constructions with line segments. By extension, these same processes apply to rays and lines; you can extend line segments and add arrows however you want.

Reproducing (Duplicating) a Line Segment

Imagine a line segment with end points P and Q as shown in Fig. 5-7A. Suppose that you want to create another line segment having the same length as PQ. First, construct a "working segment" somewhat longer than PQ. Then place a

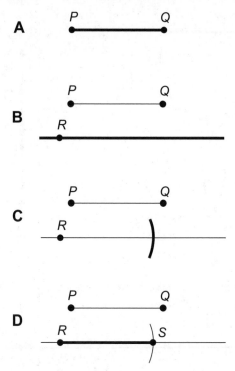

FIGURE 5-7 · Reproduction (duplication) of a line segment.

point on this "working segment" and call it *R*, as shown in Fig. 5-7B. Next, take the compass and set down the nonmarking tip on point *P*, and adjust the compass spread so that the tip of the pencil lands on point *Q* to define the length of line segment *PQ*. Now place the nonmarking tip of the compass down on point *R* and create a small arc that intersects your "working segment" as shown in Fig. 5-7C. You can define the intersection of the "working segment" and the arc as point *S* (Fig. 5-7D). The length of line segment *RS* equals that of line segment *PQ*.

Bisecting a Line Segment

Suppose that you have a line segment *PQ* (Fig. 5-8A) and you want to find the point at its center—that is, the point that bisects line segment *PQ*. First, construct an arc centered at point *P*. Make sure that the arc comprises roughly a half circle, and set the compass to span somewhat more than half the length of *PQ*. Then, without altering the setting of the compass, draw an arc centered at point *Q*, such that its radius equals the radius of the first arc that you drew

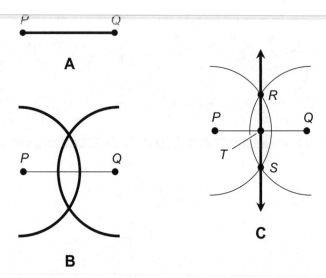

FIGURE 5-8 · Bisection of a line segment and construction of a perpendicular bisector.

(as shown in Fig. 5-8B). You can call the points at which the two arcs intersect *R* and *S*. Construct a line passing through both *R* and *S*. Under these circumstances, line *RS* intersects the original line segment *PQ* at a point *T*, which bisects line segment *PQ* (as shown in Fig. 5-8C).

Perpendicular Bisector

Imagine that you want to construct a line that bisects a specific line segment *PQ*, and that also passes through *PQ* at a right angle. Figure 5-8 shows how you can construct such a *perpendicular bisector* line (called *RS* in this example) as an "artifact" of the bisection process. The bisection process described in the previous paragraph "automatically" provides two points that lie along a perpendicular bisector.

Perpendicular Ray at a Known Point

Figure 5-9 illustrates how you can construct a perpendicular ray from a defined point *P* on a line or line segment. Begin with the scenario at Fig. 5-9A. Set the compass for a moderate span, and construct two arcs opposite each other, both centered at point *P* and both of which intersect the line or line segment. Call these intersection points *Q* and *R*, as shown in Fig. 5-9B. Increase the span of the compass, more or less doubling it (you don't have to get it exactly double).

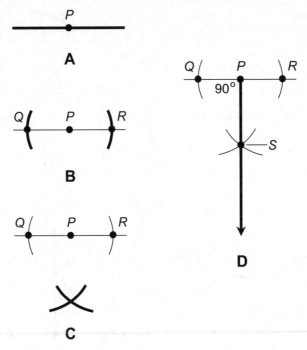

FIGURE 5-9 • Construction of a ray perpendicular to a line or line segment.

Construct an arc centered at Q and another arc centered at R, so that the two arcs have the same radius and intersect as shown in Fig. 5-9C. Now use your straight edge to draw a ray whose originating (or "back-end") point lies at P, and that passes through the intersection point (call it S) of the two arcs you just made (Fig. 5-9D). In this situation, ray PS runs outward from P at a right angle from the original line or line segment.

Dropping a Perpendicular to a Line

Figure 5-10 shows how you can draw, or *drop*, a perpendicular from a defined point P to a line that doesn't pass through that point. The term *dropping a perpendicular* means that you construct a line segment, line, or ray through a point in such a way that the line, line segment or ray "comes down on" a nearby line at a right angle.

Begin with the situation shown in Fig. 5-10A. Set the compass for a span somewhat greater than the distance between P and the line. Construct an arc that passes through the line at two points. Call these points Q and R, as shown in Fig. 5-10B.

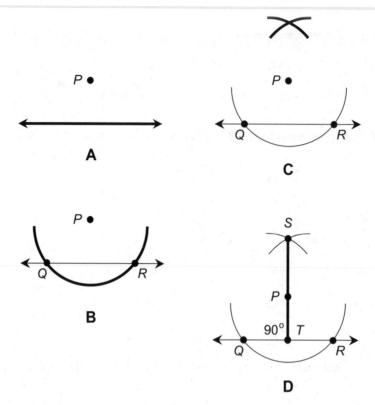

FIGURE 5-10 · Construction of a perpendicular from a point to a nearby line.

Now increase the span of the compass, roughly doubling it. (You don't have to get it exactly double.) Construct two arcs, one centered at Q and the other centered at R, such that the two arcs have the same radius and intersect each other (Fig. 5-10C). Construct a line segment that runs through P, and that also passes through the intersection point between the arcs you just made (call that point S). Extend this line segment SP until it intersects the original line. Call the resulting intersection point T, as shown in Fig. 5-10D. In this scenario, line segment PT intersects the original line at a right angle; that is, PT constitutes a perpendicular from P to the original line.

Parallel to a Line through a Specific Point

You can use several different methods to construct a parallel (line, line segment, or ray) to a specific line through a point that does not lie on that line. One of these methods takes advantage of previous constructions. Figure 5-11 portrays the process.

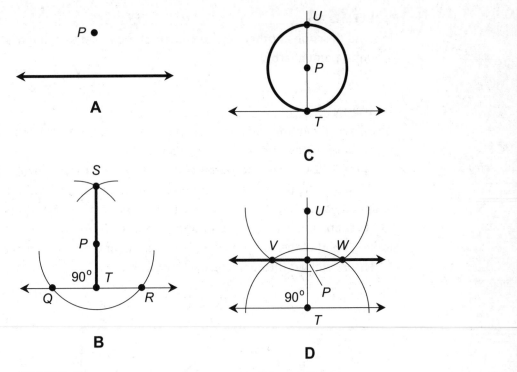

FIGURE 5-11 · Construction of a parallel through a defined point.

Suppose that you have a line segment with a point P nearby (as shown in Fig. 5-11A), and you want to create a line through P parallel to the original line. First, drop a perpendicular from P to the line using the procedure described earlier and shown in Fig. 5-10, generating points Q, R, S, and T (Fig. 5-11B). Then set the compass for the distance PT and construct a circle centered at P having a radius equal to the distance PT. This circle intersects line PT at a new point, which you can call U. Line segment UP has the same length as line segment PT (Fig. 5-11C).

Increase the span of the compass somewhat, and construct two roughly half-circular arcs having identical radii, one centered at point T and the other centered at point U, so that the arcs intersect each other at two more new points. Call these points V and W, as shown in Fig. 5-11D. In this situation, line VW runs perpendicular to line UT and also to line PT. (We know this fact because we just got done with the perpendicular construction described earlier.) Note that PT runs perpendicular to the original line. Therefore, line VW runs parallel to the original line.

TIP   *The foregoing construction provides an example in which we can correctly say, "Two perpendiculars make a parallel."*

PROBLEM 5-4

Find and describe another way to construct a parallel to a line that runs through a point nearby.

SOLUTION

The following method constitutes a scheme that can serve as a solution to this problem. (Other methods might also exist.)

Figure 5-12A shows the initial situation. Drop a perpendicular from point P to the original line, as described earlier in this chapter. This perpendicular intersects the line at point Q (Fig. 5-12B). Next, set the compass so that its span equals the length of line segment PQ. You can set the non-marking point of the compass down on point Q, and draw an arc through P to ensure you get the compass span just right.

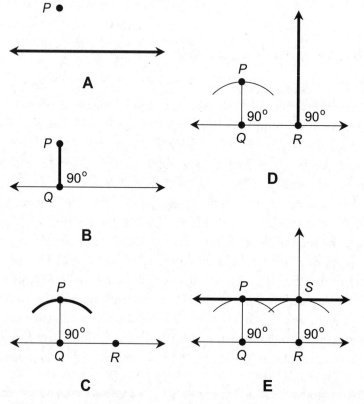

FIGURE 5-12 · Illustration for Problems 5-4 and 5-5.

Choose a second point *R* on the original line (Fig. 5-12C). Construct a perpendicular at point *R* according to the procedure described earlier in this chapter (Fig. 5-12D). Set the compass to the distance *PQ*; then place the nonmarking point of the compass on *R* and draw an arc that intersects the perpendicular. Call the intersection point *S*. You now have two points, *P* and *S*, that lie equidistant from the original line. Construct line *PS* through these points. Line *PS* runs parallel to the original line (Fig. 5-12E).

PROBLEM 5-5

Construct a square. It doesn't have to be any particular size, as long as all four sides have the same length and all four interior angles measure 90° ($\pi/2$ rad).

SOLUTION

Examine Fig. 5-12. The quadrilateral *PQRS* constitutes a rectangle, because line segments *PQ* and *RS* both run perpendicular to line *QR*, so both $\angle RQP$ and $\angle SRQ$ are right angles. You also know that lines *QR* and *PS* run parallel to each other, because that was the solution to Problem 5-4. You can therefore conclude that $\angle PSR$ and $\angle QPS$ are both right angles, because opposite interior angles to the transversals of parallel lines always have equal measure; they're congruent ($\angle RQP \cong \angle PSR$ and $\angle SRQ \cong \angle QPS$).

Now you can easily modify the construction process shown in Fig. 5-12 to ensure that the resulting quadrilateral *PQRS* constitutes a square. Instead of choosing point *R* on the original line "at random," use the compass, set so that its span equals the distance *PQ*, to determine point *R*. Set the nonmarking point of the compass down on point *Q* and draw an arc so that it intersects the original line to obtain point *R*. This action ensures that the distance *PQ* equals the distance *QR*. From there, you can complete the construction in the same way you did when you solved Problem 5-4.

Angular Construction Methods

The following paragraphs describe how to reproduce (copy or duplicate) an angle that measures less than 180° (π rad). You'll also learn how to bisect such an angle.

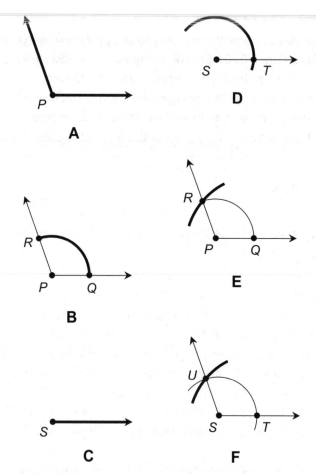

FIGURE 5-13 • Reproduction (duplication) of an angle.

Reproducing an Angle

Figure 5-13 illustrates how you can reproduce an angle. Suppose that two rays intersect at point P, as shown in Fig. 5-13A. Set down the nonmarking tip of the compass on point P and construct an arc from one ray to the other. Let R and Q represent the two points where the arc intersects the rays (Fig. 5-13B). Call the angle in question $\angle QPR$, where points R and Q lie equidistant (equally far away) from point P.

Place a new point S somewhere on the page a considerable distance away from point P and construct a ray emanating outward from point S as shown in Fig. 5-13C. This ray can run off in any direction, but you'll find that things work out best if you "send" it in approximately the same direction as ray PQ goes. Make the new ray at least as long as ray PQ. Without changing the compass span from its previous setting, place its nonmarking tip on point *S* and construct a sweeping arc that's larger than arc *QR*. You can guess at a good sweep for this arc (Fig. 5-13D), or you can make a full circle. Let point *T* represent the intersection of the new arc and the new ray.

Return to the original arc, place the nonmarking tip of the compass down on point *Q*, and construct a small arc through point *R* so that the compass spans the distance *QR* (Fig. 5-13E). Then, without changing the span of the compass, place its nonmarking tip on point *T* and construct an arc that intersects the arc centered on point *S*. Call this intersection point *U*. Finally, construct ray *SU* (Fig. 5-13F). You now have a new angle with the same measure as the original angle. That is, $\angle TSU \cong \angle QPR$.

Bisecting an Angle

Figure 5-14 illustrates a way that you can bisect an angle; that is, divide it in half. First, suppose that two rays intersect at point *P*, as shown in Fig. 5-14A. Set down the nonmarking tip of the compass on point *P* and construct an arc from one ray to the other. Let *R* and *Q* represent the two points where the arc intersects the rays (Fig. 5-14B). You can now call the angle in question $\angle QPR$, where points *R* and *Q* lie equidistant from *P*.

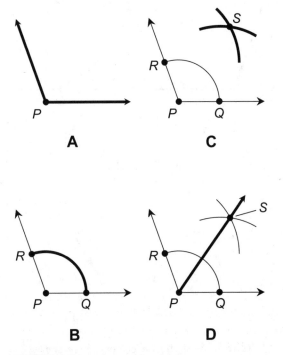

FIGURE 5-14 · Bisection of an angle.

Place the nonmarking tip of the compass on point Q, increase its span somewhat from the setting that you used to generate arc QR, and construct a new arc. Next, without changing the span of the compass, set its nonmarking tip on point R and construct an arc that intersects the arc centered on Q. If the arc centered on point Q isn't long enough, go back and make it longer. You can make it a full circle if you want. Let S represent the point at which the two arcs intersect (Fig. 5-14C). Finally, construct ray PS, as Fig. 5-14D illustrates. This ray bisects ∠QPR.

TIP   *In the foregoing construction, ∠QPS ≅ ∠SPR, and the sum of the measures of ∠QPS and ∠SPR equals the measure of ∠QPR.*

PROBLEM 5-6

Find another way to bisect an angle that measures less than 180° (π rad).

SOLUTION

Refer to Fig. 5-15. The process starts in the same way as described earlier. Two rays intersect at point P, as shown in Fig. 5-15A. Set down the

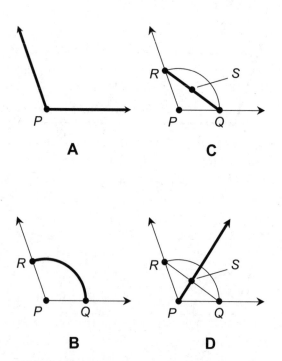

FIGURE 5-15 · Illustration for Problems 5-6 and 5-7.

nonmarking tip of the compass on point *P* and construct an arc from one ray to the other to get points *R* and *Q* (Fig. 5-15B) defining ∠*QPR*, where points *R* and *Q* lie equidistant from point *P*.

Construct line segment *RQ*. Then bisect it, following the procedure for bisecting line segments described earlier in this chapter. Call the midpoint of the line segment point *S* (Fig. 5-15C). Finally, construct ray *PS* (Fig. 5-15D). This ray bisects ∠*QPR*.

PROBLEM 5-7

Show that the angle bisection method described in the solution to Problem 5-6 works for any angle measuring less than 180° (π rad).

SOLUTION

Examine Fig. 5-15D and note the two triangles Δ*SRP* and Δ*PQS*. These triangles have corresponding sides of equal lengths:

- *SR* = *QS* (you bisected the line segment)
- *PR* = *QP* (you constructed them both from the same arc centered at *P*)
- *PS* = *PS* (any line segment has the same length as itself)

From these three facts, the side-side-side (SSS) principle from Chap. 2 assures you that Δ*SRP* and Δ*PQS* are inversely congruent. Therefore, the corresponding angles (the angles opposite corresponding sides), as you proceed around the triangles in opposite directions, have equal measure. Because *SR* = *QS*, you can conclude that ∠*SPR* and ∠*QPS* have equal measure. Because their measures obviously add up to the measure of ∠*QPR*, you know that ray *PS* bisects ∠*QPR*.

QUIZ

Refer to the text in this chapter if necessary. A good score is eight correct. Answers are in the back of the book.

1. Suppose that we want to "record" the length of a line segment for future use, such as constructing another line segment of equal length along a line. Which of the following methods constitutes a legitimate way to carry out this task?
   A. We can set a straight edge along the line segment, mark off the end points on the straight edge, and then use those marks as future reference points.
   B. We can draw circles of equal radius centered at both end points of the line segment, and then use the distance between either end point and the intersection of the circles as the "recorded" length for future reference.
   C. We can set the nonmarking point of a compass on one end point of the line segment, hold it there, and then adjust the compass span so as to place the tip of the compass pencil on the other end point of the line segment. Then we can use the two tips of the compass as future reference points.
   D. Any of the above

2. How can we construct a 45° angle?
   A. We can construct a square and then draw its diagonal. The angle between the diagonal and any one of the square's sides will equal 45°.
   B. We can construct a perpendicular bisector that intersects a line segment, and then bisect any one of the four angles between the line segment and its perpendicular bisector. Either of the resulting angles will measure 45°.
   C. We can construct a rectangle and then bisect any one of its interior angles. Either of the resulting angles will measure 45°.
   D. Any of the above

3. We must always ensure that we can complete a geometric construction
   A. without having to reproduce any angles.
   B. with lines and points only.
   C. in a finite number of steps.
   D. with only a drafting triangle and a pencil.

4. The ideal compass for performing a geometric construction
   A. has no angle-measuring scale.
   B. includes distance references along its shafts.
   C. can draw ellipses as well as circles.
   D. has two pencils, one along each shaft.

5. What's the best way to construct a line segment whose length exceeds that of your straight edge?
   A. Align the straight edge with part of the line segment and then extend the line

segment as far as you need.

B. Find a longer straight edge and then use it to construct the new line segment.

C. Use two or more identical straight edges and align them to construct the new line segment.

D. Use two or more identical drafting triangles and align them to construct the new line segment.

6. In the solution to Problem 5-1 on page 98, we learned how to construct a circle with *twice* the radius of a given circle. How can we construct a circle with *half* the radius of a given circle?

A. Draw a ray from the circle's center point out past the circle itself; then bisect the line segment connecting the center point with the point that intersects the circle; then set the compass span to the length of either half of the bisected line segment; finally draw a new circle with that radius.

B. Draw two perpendicular rays from the circle's center point out past the circle itself; then set the compass span to the distance between the points where the rays intersect the circle; finally draw a new circle with that radius.

C. Draw two perpendicular rays from the circle's center point out past the circle itself; then set the compass span to half the distance between the points where the rays intersect the circle; finally draw a new circle with that radius.

D. We can't.

7. How can we construct an angle whose measure equals $\pi/8$ rad and have complete confidence in the accuracy of our result?

A. We can construct a parallelogram and then draw its diagonal. Then we can bisect the angle between the diagonal and any one of the sides.

B. We can construct a rhombus and then draw its diagonal. Then we can bisect the angle between the diagonal and any one of the sides.

C. We can construct a rectangle and then draw its diagonal. Then we can bisect the angle between the diagonal and any one of the sides.

D. We can construct a square and then draw its diagonal. Then we can bisect the angle between the diagonal and any one of the sides.

8. How can we construct an angle whose measure equals 67.5° and have complete confidence in the accuracy of our result?

A. We can't.

B. We can construct a square, trisect any one of its interior angles, and then duplicate the result, making the new angle adjacent to the original one. The sum of these two angles will equal 67.5°.

C. We can construct an angle whose measure equals $\pi/8$ rad and then reproduce it twice, constructing the second angle adjacent to the original one and the third angle adjacent to the second one. The sum of all three angles will equal 67.5°.

D. We can bisect a straight angle (i.e., one of π rad) three times and then duplicate the result, making the new angle adjacent to any one of the angles that we got from the triple bisection. The sum of these two angles will equal 67.5°.

9. **Suppose that you want to construct a parallelogram. What should you do first?**
   A. Construct two perpendicular lines.
   B. Construct two concentric circles.
   C. Construct two parallel lines.
   D. Construct an equilateral triangle.

10. **Which of the following actions violates the formal rules for geometric construction?**
   A. Define the measure of an angle by laying a compass down on it and reading the number from a graduated scale at the compass apex.
   B. Draw a line segment by running a pencil's tip along a straight edge from one defined point to another defined point.
   C. Create a "random" angle by using a straight edge to draw two line segments that intersect at their end points.
   D. Construct a "random" circle with a compass set to any desired span.

chapter 6

The Cartesian Plane

We can define the *Cartesian plane*, also called the *rectangular coordinate plane* or *rectangular coordinates*, by constructing two calibrated number lines that intersect at a right angle. This trick allows us to pictorially describe equations that relate one variable to another. You should have a knowledge of first-year high-school algebra before tackling this chapter.

CHAPTER OBJECTIVES

In this chapter, you will

- Graph ordered pairs as points in a coordinate system.
- Calculate the distance between two points.
- Learn the difference between a relation and a function.
- Graph simple relations and functions.
- Determine equations from graphs.
- Graphically portray solutions to pairs of equations.

Two Number Lines

Figure 6-1 illustrates the simplest possible set of rectangular coordinates. Both number lines have equal increments. On either axis, any two points corresponding to consecutive integers lie the same distance apart, no matter where on the axis we look. The two number lines intersect at their zero points. We call the horizontal number line the *x axis* and the vertical number line the *y axis*.

Ordered Pairs as Points

Figure 6-2 shows three specific points, called *P*, *Q*, and *R*, plotted on the Cartesian plane. Point *P* has coordinates (−5,−4), and point *Q* has coordinates (3,5). We'll look more closely at point *R* in a few moments.

We can denote any given point as an *ordered pair* in the form (*x,y*), determined by the numerical values at which perpendiculars from the point intersect the *x* and *y* axes. In Fig. 6-2, we see the perpendiculars as horizontal and vertical dashed lines.

The word "ordered" means that the order or sequence in which we list the numbers makes a big difference! This distinction makes an ordered pair fundamentally different from a set of two numbers, in which the order or sequence doesn't matter. The ordered pair (7,2) is not the same as the ordered pair (2,7), even though both pairs contain the same two numbers. However, the sets {7, 2} and {2, 7} are identical.

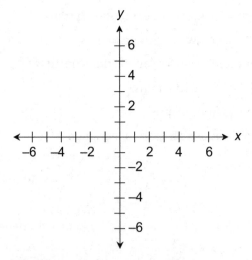

FIGURE 6-1 · A Cartesian plane contains two number lines that intersect at right angles.

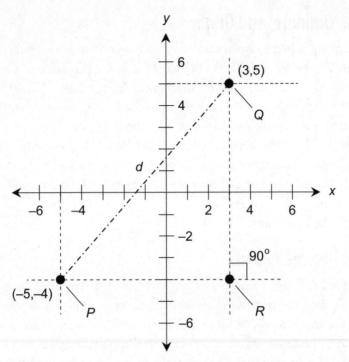

FIGURE 6-2 · Two points *P* and *Q*, plotted in rectangular coordinates, and a third point *R*, important in finding the distance *d* between *P* and *Q*.

TIP *As a matter of convention, when denoting an ordered pair, we place the two numbers or variables together right up against the comma (leaving no space after the comma). When denoting a set of two numbers, we leave a space after the comma.*

 Still Struggling

Think of a highway, which consists of a northbound lane and a southbound lane. If the highway never carries any traffic, it doesn't matter which lane (the one on the eastern side or the one on the western side) you designate as "northbound" and which lane you designate as "southbound." But once you put cars and trucks on that road, it makes a tremendous difference which direction you go in either lane! You might compare a two-element set to a two-lane road without traffic and an ordered pair to a two-lane road with traffic.

Abscissa, Ordinate, and Origin

In any graphing scheme, we always have at least one *independent variable* and at least one *dependent variable*. As the name suggests, the value of the indepen-dent variable does not "depend" on anything; it "just happens." The value of the dependent variable depends on the value of the independent variable.

We call the independent-variable coordinate (usually x) of a point on the Cartesian plane the *abscissa*. We call the dependent-variable coordinate (usually y) the *ordinate*. We call the point $(0,0)$ the *origin*. In Fig. 6-2, point P has an abscissa of –5 and an ordinate of –4, and point Q has an abscissa of 3 and an ordinate of 5. We can see, upon careful inspection, that point R has an abscissa of 3 and an ordinate of –5.

Distance between Points

Consider two different points $P = (x_0,y_0)$ and $Q = (x_1,y_1)$ on the Cartesian plane. We can calculate the distance d between these two points by determining the length of the hypotenuse, or longest side, of a right triangle PQR, where point R constitutes the intersection of a "horizontal" line through P and a "vertical" line through Q. In this case, "horizontal" means "parallel to the x axis," and "vertical" means "parallel to the y axis." Figure 6-2 shows an example.

Alternatively, we can use a "horizontal" line through Q and a "vertical" line through P to get the point R. In this case, the resulting right triangle has the same hypotenuse (line segment PQ) as the triangle determined as shown in Fig. 6-2.

Think back to Chap. 2 for a minute. Recall the Pythagorean theorem, which states that the square of the length of the hypotenuse of a right triangle equals the sum of the squares of the lengths of the other two sides. In this case, the theorem tells us that

$$d^2 = (x_1 - x_0)^2 + (y_1 - y_0)^2$$

and therefore that

$$d = [(x_1 - x_0)^2 + (y_1 - y_0)^2]^{1/2}$$

where the 1/2 power represents the square root. In the situation of Fig. 6-2, we can calculate the distance d between points $P = (x_0,y_0) = (-5,-4)$ and $Q = (x_1,y_1) = (3,5)$ as follows:

$$d = \{[3 - (-5)]^2 + [5 - (-4)]^2\}^{1/2}$$
$$= [(3 + 5)^2 + (5 + 4)^2]^{1/2}$$

$$= (8^2 + 9^2)^{1/2}$$
$$= (64 + 81)^{1/2}$$
$$= 145^{1/2}$$
$$= 12.0416 \text{ (approx.)}$$

This result is accurate to four decimal places, as determined using a standard digital calculator that can find square roots. (We assume that the coordinate values for points P and Q in Fig. 6-2 are mathematically exact.)

Relation versus Function

Let's compare the idea of a *relation* and the idea of a *function* as they pertain to coordinate geometry. A relation constitutes an equation or formula that "relates" the value of one variable to that of another. A function is a relation that meets certain specific requirements. All functions constitute relations, but not all relations constitute functions.

Relations

We can denote a relation between two variables x and y so that it expresses the value of y in terms of the value of x. In this format, y represents the dependent variable and x represents the independent variable. Some examples follow:

$$y = 5$$
$$y = x + 1$$
$$y = 2x$$
$$y = x^2$$

Some Simple Graphs

Figure 6-3 shows how the graphs of the above equations look on the Cartesian plane. Mathematicians and scientists call such a graph a *curve*, even if it happens to be a straight line.

The graph of $y = 5$ (curve A) appears as a horizontal line passing through the point $(0,5)$ on the y axis. The graph of $y = x + 1$ (curve B) is a straight line that ramps upward at a 45° angle (from left to right) and passes through $(0,1)$ on the y axis. The graph of $y = 2x$ (curve C) shows up as a straight line that ramps upward more steeply, and that passes through the origin $(0,0)$. The graph of

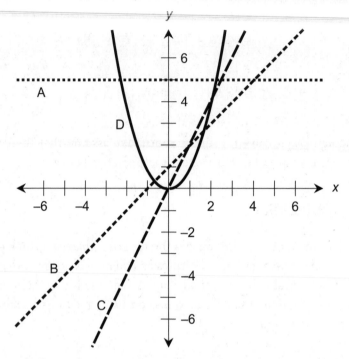

FIGURE 6-3 · Graphs of four relations in Cartesian coordinates. Drawings A, B, and C show linear relations; drawing D portrays a nonlinear relation.

$y = x^2$ (curve D) appears as a geometric curve called a *parabola*. In this case, the parabola rests on the origin $(0,0)$, opens upward, and exhibits left-to-right (*bilateral*) symmetry with respect to the y axis.

TIP   *In Fig. 6-3, graphs A, B, and C portray so-called* **linear relations** *because they appear as straight lines in the Cartesian coordinate plane. Graph D portrays a* **nonlinear relation** *because it does not appear as a straight line in the Cartesian plane.*

Functions

All of the relations shown in Fig. 6-3 share a feature that we can identify by examining their graphs: For every abscissa, each relation contains *at most* one ordinate. Never does a curve have two or more ordinates for a single abscissa, although one of them (the parabola, curve D) has two abscissas for all positive ordinates.

We can define a function as a mathematical relation in which every abscissa corresponds to *at most* one ordinate. According to this criterion, all four of the curves shown in Fig. 6-3 portray functions of y in terms of x. In addition, curves A, B, and C show functions of x in terms of y. But curve D does not represent a function of x in terms of y. If we consider x as the dependent variable and y as the independent variable, then there exist some values of y (some abscissas) that "mate" with two values of x (ordinates).

Let's denote functions as italicized letters of the alphabet such as f, F, g, G, h, or H, followed by the independent variable in parentheses. Consider these examples:

$$f(x) = 5$$
$$g(x) = x + 1$$
$$h(x) = 2x$$
$$F(x) = x^2$$

We can read these equations out loud as "f of x equals 5," "g of x equals x plus 1," "h of x equals 2 times x," and "F of x equals x squared," respectively.

PROBLEM 6-1

Plot the following points on the Cartesian plane: (–2,3), (3,–1), (0,5), and (–3,–3).

SOLUTION

Figure 6-4 shows these points. The dashed lines are perpendiculars, dropped to the axes to show the x and y coordinates of each point for reference purposes only. (The actual graphs of the points do not include these dashed lines.)

PROBLEM 6-2

What's the distance between the two points (0,5) and (–3,–3) in Fig. 6-4? Express the answer to three decimal places. Assume that the coordinate values are exact.

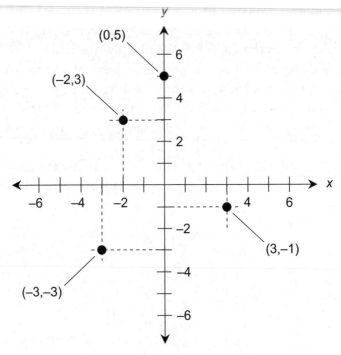

FIGURE 6-4 • Illustration for Problems 6-1 and 6-2.

 SOLUTION

Let's say that $(x_0,y_0) = (0,5)$ and $(x_1,y_1) = (-3,-3)$. We calculate the distance d between these two points as follows:

$$d = [(x_1 - x_0)^2 + (y_1 - y_0)^2]^{1/2}$$
$$= [(-3 - 0)^2 + (-3 - 5)^2]^{1/2}$$
$$= [(-3)^2 + (-8)^2]^{1/2}$$
$$= (9 + 64)^{1/2}$$
$$= 73^{1/2}$$
$$= 8.544 \text{ (rounded off)}$$

Straight Lines

We can always represent a straight line on the Cartesian plane as a *linear equation*. Several different forms exist for linear equations. No matter what form a linear equation shows up in at first, we use algebra to "morph" it into an equation where neither x nor y is raised to any power other than 0 or 1.

Standard Form of Linear Equation

The *standard form of a linear equation* in variables x and y comprises constant multiples of the two variables, plus another constant, all summed up to equal zero, as follows:

$$ax + by + c = 0$$

In this "generic" equation, we denote the *constants* as a, b, and c. If a constant happens to equal 0, then we don't have to write it down, nor do we have to write its multiple (by either x or y). Examples of linear equations in the standard form include the following:

$$2x + 5y - 3 = 0$$
$$5y - 3 = 0$$
$$2x - 3 = 0$$
$$2x = 0$$
$$5y = 0$$

Still Struggling

You can divide each side of the fourth (next-to-last) of the above equations by 2, thereby simplifying it to $x = 0$. Similarly, you can divide each side of the fifth (last) equation by 5, simplifying it to $y = 0$.

Slope-Intercept Form of Linear Equation

We can manipulate any linear equation in variables x and y to make it easy to plot on the Cartesian plane. We can convert a linear equation from standard form to *slope-intercept form* by going through several steps. Let's start with the general equation

$$ax + by + c = 0$$

Subtracting c from each side, we get

$$ax + by = -c$$

We can subtract ax from each side to obtain

$$by = -ax - c$$

When we divide through by b, we get

$$y = (-a/b)x - c/b$$

We can also express this equation as

$$y = (-a/b)x + (-c/b)$$

where a, b, and c represent real-number constants, and $b \neq 0$. We call the quantity $-a/b$ the *slope* of the line (also known as "rise over run"), an indicator of how steeply and in what sense the line slants. The quantity $-c/b$ represents the ordinate (or y-value) of the point at which the line crosses the y axis; we call it the *y-intercept*.

Definition of Slope

Suppose that dx represents a small change in the value of x on the graph of a line. Let dy represent the change in the value of y that results from this change in x. We define the ratio dy/dx as the slope of the line. Let's symbolize the slope as m. Now imagine that some number k represents the y-intercept for the line. We can derive m and k from a, b, and c in the above-defined equation as follows, provided that $b \neq 0$:

$$m = -a/b$$

and

$$k = -c/b$$

We can rewrite the linear equation in slope-intercept form as

$$y = (-a/b)x + (-c/b)$$

Substituting m for $-a/b$ and k for $-c/b$, we get

$$y = mx + k$$

Plotting the Line

When you want to plot the graph of a linear equation in Cartesian coordinates, proceed as follows:

- Convert the equation to slope-intercept form.
- Plot the point $y = k$.

- Move to the right by n units on the graph, where n is some number that represents some reasonable distance on the graph.
- If m is positive, move upward mn units.
- If m is negative, move downward $|m|n$ units, where $|m|$ equals the absolute value of m.
- If $m = 0$, don't move up or down at all.
- Plot the resulting point.
- Connect the two points with a straight line.

Figures 6-5A and 6-5B illustrate the graphs of two different linear relations in slope-intercept form. At A, we see the graph of the equation

$$y = 5x - 3$$

At B, we see the graph of the equation

$$y = -x + 2$$

Still Struggling

Positive slope indicates that a line ramps upward as you move from left to right, and negative slope indicates that a line ramps downward as you move from left to right. A slope of 0 indicates a horizontal line. We can't define the slope of a vertical line because, in the form we've learned here, a vertical line requires that m consist of a quotient with a denominator equal to 0.

Point-Slope Form of Linear Equation

We'll sometimes have trouble plotting the graph of a line based on the y-intercept (the point at which the line intersects the y axis) when the part of the graph of interest lies far from the y axis. In this sort of situation, we can use the *point-slope form* of a linear equation to help us draw the graph. We can express a line in this form if we know the slope m of the line and the coordinates of a known point (x_0, y_0), as follows:

$$y - y_0 = m(x - x_0)$$

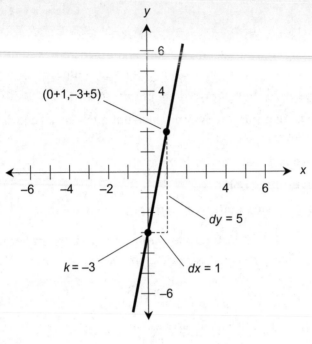

A

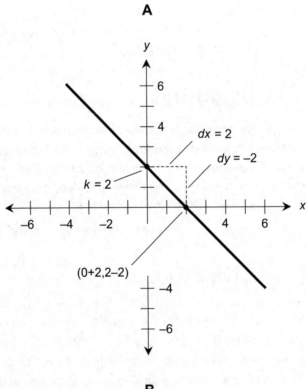

B

FIGURE 6-5 · A. Graph of the linear equation $y = 5x - 3$. **B.** Graph
of the linear equation $y = -x + 2$.

When we want to plot a graph of a linear equation using the point-slope method, we can follow these steps in order:

- Convert the equation to point-slope form.
- Determine a point (x_0, y_0) by "plugging in" values.
- Plot (x_0, y_0) on the coordinate plane.
- Move to the right by n units on the graph, where n is some number that represents a reasonable distance on the graph.
- If m is positive, move upward mn units.
- If m is negative, move downward $|m|n$ units, where $|m|$ equals the absolute value of m.
- If $m = 0$, don't move up or down at all.
- Plot the resulting point (x_1, y_1).
- Connect the points (x_0, y_0) and (x_1, y_1) with a straight line.

Figure 6-6A shows the graph of the following linear equation based on the point-slope form:

$$y - 104 = 3(x - 72)$$

Figure 6-6B portrays the graph of another linear equation based on the point-slope form:

$$y + 55 = -2(x + 85)$$

Finding Linear Equation Based on Graph

Imagine that we're working in the Cartesian plane, and we know the exact coordinates of two distinct points P and Q. These two points, no matter where they lie, define a unique straight line. Let's call the line L and give the coordinates of the points the names

$$P = (x_p, y_p)$$

and

$$Q = (x_q, y_q)$$

We can calculate the slope m of line L using either of the following formulas:

$$m = (y_q - y_p)/(x_q - x_p)$$

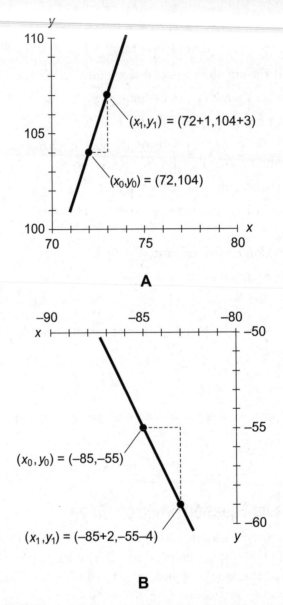

FIGURE 6-6 · A. Graph of the linear equation $y - 104 = 3(x - 72)$.
B. Graph of the linear equation $y + 55 = -2(x + 85)$.

or

$$m = (y_p - y_q)/(x_p - x_q)$$

provided that $x_p \neq x_q$. (If $x_p = x_q$, we get denominators of 0 in the formulas, preventing us from defining the slope.)

We can determine the point-slope equation of the line L based on the known coordinates of P or Q. Either of the following formulas represent L:

$$y - y_p = m(x - x_p)$$

or

$$y - y_q = m(x - x_q)$$

Parabolas and Circles

The Cartesian-coordinate graph of a *quadratic equation* always shows up as a parabola. We can write down any quadratic equation in the general form

$$y = ax^2 + bx + c$$

where a, b, and c represent real-number constants, and $a \neq 0$. (If $a = 0$, then we have a linear equation, not a quadratic equation.)

When we want to plot a graph of an equation that appears in the above form, we first determine the coordinates of the following point (x_0, y_0), as follows:

$$x_0 = -b/(2a)$$

and

$$y_0 = c - b^2/(4a)$$

The coordinates (x_0, y_0) define the *vertex point* of the parabola. That's the point at which the curvature is sharpest, and at which a line tangent to (i.e., a line that "brushes up against") the curve runs horizontally. For the Cartesian graph of a quadratic equation that tells us y in terms of x, we can have either of two cases:

- In a parabola that opens straight upward, the vertex is the graph's minimum.
- In a parabola that opens straight downward, the vertex is the graph's maximum.

Once we know the vertex, we can find four more points by "plugging in" values of x somewhat greater than and less than x_0 and then determining the corresponding y-values. Let's call these x-values by the names x_{-2}, x_{-1}, x_1, and x_2. We should space them evenly on either side of x_0, such that

$$x_{-2} < x_{-1} < x_0 < x_1 < x_2$$

and

$$x_{-1} - x_{-2} = x_0 - x_{-1} = x_1 - x_0 = x_2 - x_1$$

This arrangement produces five points that lie along the parabola, and that exhibit symmetry relative to the axis of the curve. We can now fill in the graph if we've wisely chosen the points. If $a > 0$, the parabola opens upward. If $a < 0$, the parabola opens downward.

Plotting a Parabola

Consider the following equation for y in terms of x:

$$y = x^2 + 2x + 1$$

This equation has coefficients of $a = 1$, $b = 2$, and $c = 1$. Using the formula defined above, we can calculate the x-value of the vertex point as

$$x_0 = -b/(2a)$$
$$= -2/(2 \times 1)$$
$$= -2/2$$
$$= -1$$

and we can calculate the y-value of the vertex point as

$$y_0 = c - b^2/(4a)$$
$$= 1 - 2^2/(4 \times 1)$$
$$= 1 - 4/4$$
$$= 1 - 1$$
$$= 0$$

Therefore, we can express our first point as

$$(x_0, y_0) = (-1, 0)$$

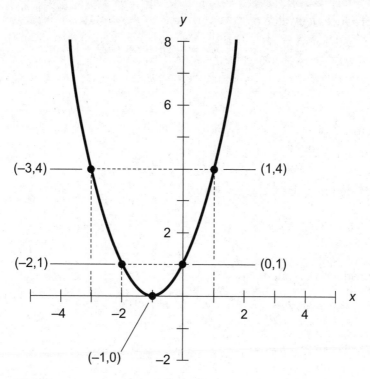

FIGURE 6-7 · Graph of the quadratic equation $y = x^2 + 2x + 1$.

Figure 6-7 illustrates this situation. Next, let's plot the points corresponding to x_{-2}, x_{-1}, x_1, and x_2, spaced at 1-unit intervals on either side of x_0. First, we define x_{-2} as

$$x_{-2} = x_0 - 2$$
$$= -3$$

which produces a y-value of

$$y_{-2} = (-3)^2 + 2 \times (-3) + 1$$
$$= 9 - 6 + 1$$
$$= 4$$

so therefore

$$(x_{-2}, y_{-2}) = (-3, 4)$$

Next, we define x_{-1} as

$$x_{-1} = x_0 - 1$$
$$= -2$$

which produces a y-value of

$$y_{-1} = (-2)^2 + 2 \times (-2) + 1$$
$$= 4 - 4 + 1$$
$$= 1$$

so therefore

$$(x_{-1}, y_{-1}) = (-2, 1)$$

Next, we define x_1 as

$$x_1 = x_0 + 1$$
$$= 0$$

which produces a y-value of

$$y_1 = 0^2 + 2 \times 0 + 1$$
$$= 0 + 0 + 1$$
$$= 1$$

so therefore

$$(x_1, y_1) = (0, 1)$$

Finally, we define x_2 as

$$x_2 = x_0 + 2$$
$$= 1$$

which produces a y-value of

$$y_2 = 1^2 + 2 \times 1 + 1$$
$$= 1 + 2 + 1$$
$$= 4$$

so therefore

$$(x_2, y_2) = (1, 4)$$

When we draw these five points on the Cartesian plane, we get a good idea of where the parabola lies, allowing us to easily "fill in the curve" as shown in Fig. 6-7.

Plotting Another Parabola

Let's try another example, this time with a parabola that opens downward instead of upward. Consider the equation

$$y = -2x^2 + 4x - 5$$

This equation has coefficients of $a = -2$, $b = 4$, and $c = -5$. Using the formula defined a little while ago, we can calculate the x-value of the vertex point as

$$x_0 = -b/(2a)$$
$$= -4/[2 \times (-2)]$$
$$= -4/(-4)$$
$$= 1$$

and the y-value of the vertex point as

$$y_0 = c - b^2/(4a)$$
$$= -5 - 4^2/[4 \times (-2)]$$
$$= -5 - 16/(-8)$$
$$= -5 + 2$$
$$= -3$$

so therefore

$$(x_0, y_0) = (1, -3)$$

We plot this point first, as shown in Fig. 6-8. Now we're ready to plot the points corresponding to x_{-2}, x_{-1}, x_1, and x_2, spaced at 1-unit intervals on either side of x_0. First, we define x_{-2} as

$$x_{-2} = x_0 - 2$$
$$= -1$$

which produces a y-value of

$$y_{-2} = -2 \times (-1)^2 + 4 \times (-1) - 5$$
$$= -2 - 4 - 5$$
$$= -11$$

so therefore

$$(x_{-2}, y_{-2}) = (-1, -11)$$

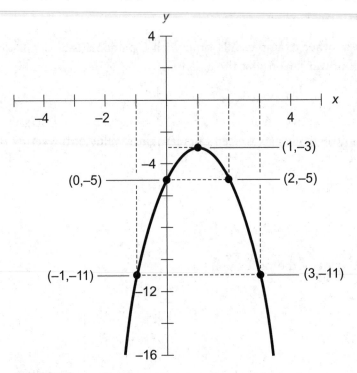

FIGURE 6-8 · Graph of the quadratic equation $y = -2x^2 + 4x - 5$.

Next, we define x_{-1} as

$$x_{-1} = x_0 - 1$$
$$= 0$$

which produces a y-value of

$$y_{-1} = -2 \times 0^2 + 4 \times 0 - 5$$
$$= -5$$

so therefore

$$(x_{-1}, y_{-1}) = (0, -5)$$

Next, we define x_1 as

$$x_1 = x_0 + 1$$
$$= 2$$

which produces a y-value of

$$y_1 = -2 \times 2^2 + 4 \times 2 - 5$$
$$= -8 + 8 - 5$$
$$= -5$$

so therefore

$$(x_1, y_1) = (2, -5)$$

Finally, we define x_2 as

$$x_2 = x_0 + 2$$
$$= 3$$

which produces a y-value of

$$y_2 = -2 \times 3^2 + 4 \times 3 - 5$$
$$= -18 + 12 - 5$$
$$= -11$$

so therefore

$$(x_2, y_2) = (3, -11)$$

Now that we know five distinct points that fall in "good places" on the curve, we can draw the parabola by "connecting the dots."

Equation of Circle

The general form for the equation of a *circle* in the xy-plane shows symmetry with respect to both variables (just as a circle has symmetry in both the horizontal sense and the vertical sense). We have

$$(x - x_0)^2 + (y - y_0)^2 = r^2$$

where (x_0, y_0) represents the coordinates of the center of the circle, and r represents the circle's *radius*, or distance from the center to any point on the curve itself. Figure 6-9 illustrates a generic example. In the special case where the circle's center lies at the origin $(0,0)$ of the Cartesian plane, the formula simplifies to

$$x^2 + y^2 = r^2$$

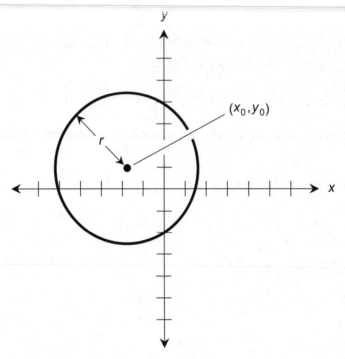

FIGURE 6-9 • Circle centered at (x_0, y_0) with radius r.

Such a circle intersects the x axis at the points $(r,0)$ and $(-r,0)$; it intersects the y axis at the points $(0,r)$ and $(0,-r)$. An even more specific case is the *unit circle*. We can express it in terms of the formula

$$x^2 + y^2 = 1$$

This curve intersects the x axis at the points $(1,0)$ and $(-1,0)$; it intersects the y axis at the points $(0,1)$ and $(0,-1)$.

 PROBLEM 6-3

Draw a Cartesian graph of the circle represented by $(x - 1)^2 + (y + 2)^2 = 9$.

SOLUTION

Based on the general formula for a circle, we can determine that the center point has coordinates $x_0 = 1$ and $y_0 = -2$. The radius equals the square root of 9, which equals 3. We therefore have a circle whose center point lies at $(1,-2)$ on the Cartesian plane, and whose radius equals 3 units as shown in Fig. 6-10.

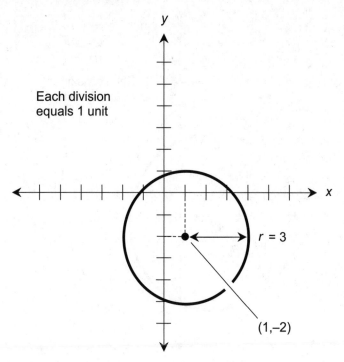

Each division
equals 1 unit

$r = 3$

$(1,-2)$

FIGURE 6-10 · Illustration for Problem 6-3.

PROBLEM 6-4

Determine the equation of the circle graphed in Fig. 6-11.

SOLUTION

First, let's note that the center point has coordinates $(-8,-7)$, so we can assign it the coordinate values

$$x_0 = -8$$

and

$$y_0 = -7$$

The radius r equals 20. When we square it, we get

$$r^2 = 20 \times 20$$

$$= 400$$

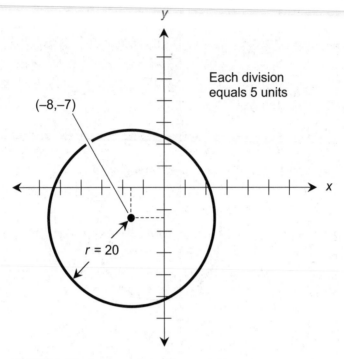

FIGURE 6-11 · Illustration for Problem 6-4.

We recall that the general formula for a circle in Cartesian coordinates is

$$(x - x_0)^2 + (y - y_0)^2 = r^2$$

Inputting our known values, we get

$$[x - (-8)]^2 + [y - (-7)]^2 = 400$$

which simplifies to

$$(x + 8)^2 + (y + 7)^2 = 400$$

Solving Pairs of Equations

We can envision and approximate the solutions to pairs of equations by graphing both of the equations on the same coordinate grid. Solutions appear as intersection points between the graphs.

A Line and a Curve

Suppose that you encounter two equations in the two variables x and y. You want to determine the values of x and y (if any) that satisfy both equations. In this scenario, you have a pair of so-called *simultaneous equations*. Consider the following example:

$$y = x^2 + 2x + 1$$

and

$$y = -x + 1$$

Figure 6-12 portrays the graphs of these equations. The graph of the first equation appears as a parabola (solid curve), and the graph of the second equation shows up as a straight line (dashed). The line crosses the parabola at two points, indicating that two real-number solutions exist for this pair of simultaneous

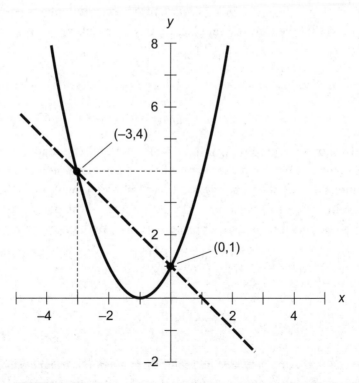

FIGURE 6-12 · Graphs of two equations, showing solutions as intersection points.

equations. We can estimate the coordinates of the points by examining the graph. It appears that they're close to, or maybe exactly,

$$(x_1,y_1) = (-3,4)$$

and

$$(x_2,y_2) = (0,1)$$

TIP *If you've taken an algebra course that taught you how to solve pairs of simultaneous equations, you can use that knowledge here and calculate the solutions to the above equations exactly. If your algebra course didn't get that far, you can nevertheless check out the above stated solutions and verify that they're exact! Just "plug in" the solutions to both equations and grind out the arithmetic.*

Another Line and Curve

Consider another pair of *two-by-two equations* (two simultaneous equations in two variables) that we can solve approximately by graphing

$$y = -2x^2 + 4x - 5$$

and

$$y = -2x - 5$$

Figure 6-13 shows the graphs. Again, the graph of the first equation constitutes a parabola (solid curve), and the graph of the second equation shows up as a straight line (dashed). The line crosses the parabola at two points, indicating that two real-number solutions exist. The coordinates of the points, corresponding to the solutions, appear to be approximately, or perhaps exactly,

$$(x_1,y_1) = (3,-11)$$

and

$$(x_2,y_2) = (0,-5)$$

TIP *Again, if you want, go ahead and solve these equations using algebra, and find the values exactly. Alternatively, you can input the above stated solutions and use simple arithmetic to verify that they're exact.*

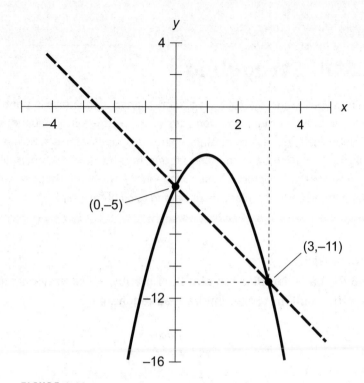

FIGURE 6-13 • Another example of equation solutions shown as the intersection points of their graphs.

Multiple Solutions

Graphing simultaneous equations can reveal general facts about them, but we can't rely on graphs to provide us with exact solutions. In real-life scientific applications, graphs rarely show us exact solutions unless they're so labeled and represent theoretical ideals.

A Cartesian-coordinate graph with real-number axes can reveal that a pair of equations has two or more real-number solutions, or only one real-number solution, or no real-number solutions at all. The real-number solutions to pairs of equations always show up as intersection points on their graphs. Therefore, if *n* intersection points exist between the curves representing two equations, then the pair of equations has *n* real-number solutions.

If a pair of equations is complicated, or if the graphs portray the results of experiments, we'll occasionally run into situations where we can't use algebra to solve them. Then graphs, with the aid of computer programs to closely approximate the points of intersection between graphs, offer the only practical means of solving simultaneous equations.

Still Struggling

Sometimes you'll want to see if a set of more than two equations in x and y has any solutions. One or more *equation pairs* within a large set of equations may have solutions; they show up as points where two graphs intersect. However, it's unusual for a set of three or more equations in x and y to have any solutions when considered *all together* (i.e., simultaneously). For that to happen, at least one point in the Cartesian plane must belong to *all* of the graphs.

 **PROBLEM 6-5**

Using the Cartesian plane to plot their graphs, we can say certain things about the solutions to the simultaneous equations

$$y = x + 3$$

and

$$(x - 1)^2 + (y + 2)^2 = 9$$

What can we say, specifically?

SOLUTION

Figure 6-14 shows the graphs of these equations. The first equation graphs as a straight line (dashed), ramping up toward the right with slope equal to 1 and intersecting the y axis at (0,3). The second equation graphs as a circle (solid curve) whose radius equals 3 units, and that's centered at the point (1,–2). We can see that this line and circle do not intersect anywhere in the Cartesian plane, so we know that there exist no real-number solutions to this pair of simultaneous equations.

 PROBLEM 6-6

Using the Cartesian plane to plot their graphs, we can say certain things about the solutions to the simultaneous equations

$$y = 1$$

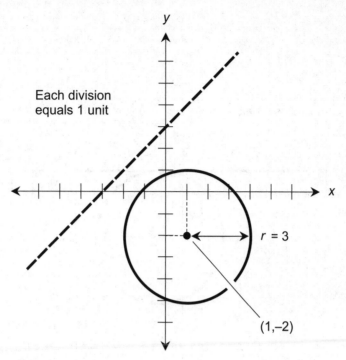

Each division
equals 1 unit

$r = 3$

$(1,-2)$

FIGURE 6-14 · Illustration for Problem 6-5.

and

$$(x - 1)^2 + (y + 2)^2 = 9$$

What can we say, specifically?

✔**SOLUTION**

Figure 6-15 shows the graphs. The first equation graphs as a horizontal straight line (dashed) intersecting the y axis at (0,1). The second equation graphs as a circle (solid curve) whose radius equals 3 units, centered at the point (1,–2). It appears from the graph that the equations have a single common solution denoted by the point (1,1), indicating that $x = 1$ and $y = 1$.

Let's use algebra to solve the equations and find out if the graph tells us the true story. Substituting 1 for y in the equation of a circle (because one of the equations tells us that $y = 1$), we get a single equation in a single variable:

$$(x - 1)^2 + (1 + 2)^2 = 9$$

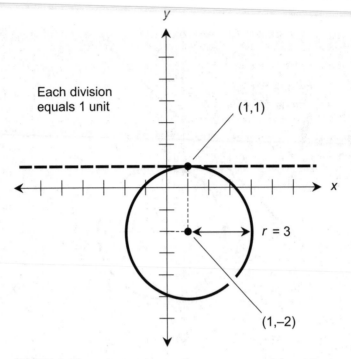

FIGURE 6-15 · Illustration for Problem 6-6.

This equation simplifies to

$$(x - 1)^2 + 3^2 = 9$$

and further to

$$(x - 1)^2 + 9 = 9$$

Subtracting 9 from each side, we get

$$(x - 1)^2 = 0$$

When we take the square root of both sides, we obtain

$$x - 1 = 0$$

Adding 1 to each side gives us the solution

$$x = 1$$

It checks out! Now we know that there exists only one solution to this pair of simultaneous equations: $x = 1$ and $y = 1$, denoted by the point $(1,1)$.

QUIZ

Refer to the text in this chapter if necessary. A good score is eight correct. Answers are in the back of the book.

1. **How far from the origin does point P lie in Fig. 6-16?**
    A. 10 units
    B. The square root of 10 units
    C. 7 units
    D. The square root of 29 units

2. **How far from the origin does point Q lie in Fig. 6-16?**
    A. 5 units
    B. The square root of 10 units
    C. The square root of 50 units
    D. 7 units

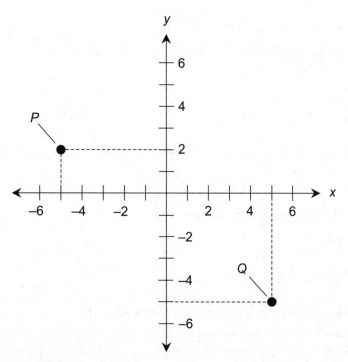

FIGURE 6-16 • Illustration for Quiz Questions 1 through 5.

3. What's the distance between points *P* and *Q* in Fig. 6-16?
   A. The square root of 17 units
   B. The square root of 79 units
   C. The square root of 149 units
   D. 13 units

4. Suppose that we draw a straight line passing through both points *P* and *Q* in Fig. 6-16. What's the slope of that line?
   A. −2/5
   B. −7/10
   C. 5/2
   D. 10/7

5. Which of the following expressions constitutes a point-slope equation for a straight line passing through points *P* and *Q* in Fig. 6-16?
   A. $y - 2 = (-7/10)(x + 5)$
   B. $y + 5 = (5/2)(x - 2)$
   C. $y + 2 = (-2/5)(x + 5)$
   D. $y - 5 = (10/7)(x + 2)$

6. Consider a parabola represented by the following equation in Cartesian coordinates:

$$y = -2x^2 + 8x - 3$$

   What are the coordinates of the parabola's vertex point?
   A. (7,1)
   B. (2,5)
   C. (8,−3)
   D. (−3/2,4)

7. Consider a circle represented by the following equation in Cartesian coordinates:

$$(x + 2)^2 + (y - 7)^2 = 196$$

   What are the coordinates of the circle's center?
   A. (7,−2)
   B. (−7,2)
   C. (−2,7)
   D. (2,−7)

8. What's the radius of the circle described in Question 7?
   A. 196 units
   B. 56 units
   C. 28 units
   D. 14 units

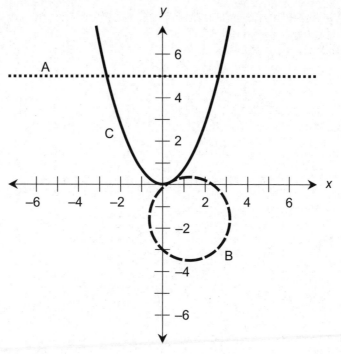

FIGURE 6-17 · Illustration for Quiz Questions 9 and 10.

9. Figure 6-17 shows graphs of a linear equation (line A), a circular equation (curve B), and a quadratic equation (curve C). Based on the appearance of the graphs, if we undertake to solve the equations for line A and curve B simultaneously, we should expect to get

   A. no real-number solutions.
   B. one real-number solution.
   C. two real-number solutions.
   D. infinitely many real-number solutions.

10. Based on the appearance of the graphs in Fig. 6-17, if we undertake to solve the equations for curves B and C simultaneously, we should expect to get

    A. no real-number solutions.
    B. one real-number solution.
    C. two real-number solutions.
    D. infinitely many real-number solutions.

Test: Part I

Do not refer to the text when taking this test. You may draw diagrams or use a calculator if necessary. A good score is at least 38 correct. Answers are in the back of the book. It's best to have a friend check your score the first time, so you won't memorize the answers if you want to take the test again.

1. When we encounter a plane polygon in which none of the vertices "bend inward" (i.e., where every interior angle measures less than 180°), we call the polygon

   A. acute.
   B. amorphous.
   C. disjoint.
   D. regular.
   E. convex.

2. Imagine two triangles, both of which have equal base lengths and equal heights. Based on this information, we can have complete confidence that the two triangles

   A. exhibit direct similarity.
   B. have equal interior areas.
   C. have equal perimeters.
   D. exhibit inverse congruence.
   E. None of the above

3. When we encounter a plane polygon whose sides all measure the same length and interior angles all have the same measure, we call the figure

   A. obtuse.
   B. polymorphous.
   C. regular.
   D. disjoint.
   E. amorphous.

4. In Fig. Test I-1, line *M* constitutes

   A. a parallel bisector of line segment *PR*.
   B. a perpendicular bisector of line segment *PR*.

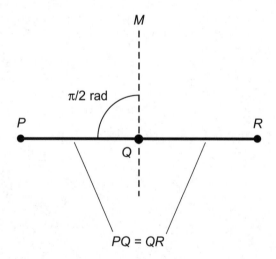

FIGURE TEST I-1 • Illustration for Part I Test Questions 4 and 5.

C. an acute bisector of line segment *PR*.
D. an obtuse bisector of line segment *PR*.
E. a radial bisector of line segment *PR*.

5. According to the appearance of Fig. Test I-1, we can surmise that line segment *PR* is
    A. closed.
    B. half-open.
    C. open.
    D. infinite.
    E. congruent.

6. Suppose that we *circumscribe* a circle with a regular polygon having *n* sides (where *n* represents a positive integer larger than 3), and then we increase *n* without limit, all the while making sure that the polygon keeps on "snugly" circumscribing the circle. As we carry out this action, the measures of the polygon's individual interior angles approach
    A. $\pi/3$ rad.
    B. $\pi/2$ rad.
    C. $2\pi/3$ rad.
    D. π rad.
    E. 2π rad.

7. A full circular revolution yields an angular measure of
    A. $\pi/4$ rad.
    B. $\pi/2$ rad.
    C. π rad.
    D. 2π rad.
    E. 4π rad.

8. Suppose that we *inscribe* a circle with a regular polygon having *n* sides (where *n* represents a positive integer larger than 3), and then we increase *n* without limit, all the while making sure that the polygon keeps on "snugly" inscribing the circle. As we carry out this action, the measures of the polygon's individual interior angles approach
    A. $\pi/3$ rad.
    B. $\pi/2$ rad.
    C. $2\pi/3$ rad.
    D. π rad.
    E. 2π rad.

9. When you use a drafting compass and straight edge to perform a geometric construction, you must *never*
    A. use either instrument more than once.
    B. use your pencil all by itself to define a point.
    C. use calibrated scales on either instrument.
    D. draw circles of arbitrary radius.
    E. draw line segments of arbitrary length.

10. Two angles in the same plane complement each other if and only if the sum of their measures equals

   A. 180°.
   B. a full circle.
   C. $\pi/2$ rad.
   D. half of a full circle.
   E. π rad.

11. The triangles illustrated in Fig. Test I-2 are both

   A. acute.
   B. isosceles.
   C. equilateral.
   D. All of the above
   E. None of the above

12. We can have absolute confidence that the triangles shown in Fig. Test I-2 exhibit one, and only one, of the following properties. Which one?

   A. Direct congruence
   B. Inverse similarity
   C. Inverse congruence
   D. The sum of all the angular measures equals $\pi/2$ rad.
   E. They both have the same perimeter.

$u = v = w$

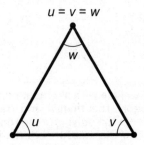

$x = y = z$

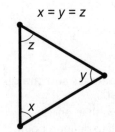

FIGURE TEST I-2 · Illustration for Part I Test Questions 11 and 12.

13. Given any three distinct points, they *cannot* form a triangle if they all lie
    A. on the same line.
    B. in a single plane.
    C. on a single rectangle.
    D. on a single circle.
    E. in a single coordinate system.

14. Consider the following equation that represents a straight line in Cartesian coordinates:

$$y = 2x - 7$$

What's the slope of this line?
    A. −2/7
    B. −7/2
    C. −7
    D. −14
    E. 2

15. In order to "qualify" as a quadrilateral, a geometric plane figure must have all of the following characteristics except one. Which one?
    A. The figure must have four distinct sides.
    B. Each interior angle must measure less than 180°.
    C. The figure must have four distinct vertex points.
    D. All the sides must have positive, finite length.
    E. All the sides must be straight line segments.

16. Imagine a triangle with interior angles measuring $\pi/4$ rad, $\pi/4$ rad, and $\pi/2$ rad. From this information, we can have complete confidence that the figure constitutes
    A. a concave triangle.
    B. a disjoint triangle.
    C. an isosceles triangle.
    D. an equilateral triangle.
    E. an obtuse triangle.

17. Consider a circle represented by the following equation in Cartesian coordinates:

$$(x + 6)^2 + (y + 3)^2 = 124$$

What are the coordinates of the center?
    A. (6,3)
    B. (−6,3)
    C. (6,−3)
    D. (−6,−3)
    E. We need more information to answer this question.

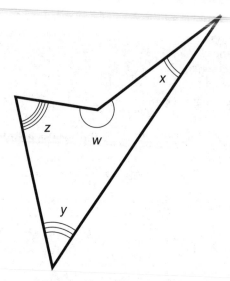

FIGURE TEST I-3 • Illustration for Part I Test
Questions 19 through 21.

18. **What's the radius of the circle described in Question 17?**

    A. 124 units
    B. The square root of 124 units
    C. 62 units
    D. 31 units
    E. We need more information to answer this question.

19. **Assuming that the entire object in Fig. Test I-3 lies in a single plane, we can surmise from its general appearance that it portrays a**

    A. rhombus.
    B. trapezoid.
    C. pentagon.
    D. parallelogram.
    E. quadrilateral.

20. **Assuming that the entire object in Fig. Test I-3 lies in a single plane, we can have complete confidence that**

    A. $w + x + y + z = \pi/2$ rad.
    B. $w + x + y + z = \pi$ rad.
    C. $w + x + y + z = 2\pi$ rad.
    D. $w + x + y + z = 3\pi$ rad.
    E. $w + x + y + z = 4\pi$ rad.

21. Assuming that the entire object in Fig. Test I-3 lies in a single plane, we can have complete confidence that each and every individual angle *w*, *x*, *y*, or *z* measures less than
    A. $\pi/4$ rad.
    B. $\pi/3$ rad.
    C. $\pi/2$ rad.
    D. π rad.
    E. 2π rad.

22. We call the independent-variable coordinate (usually *x*) of a point on the Cartesian plane the
    A. magnitude.
    B. abscissa.
    C. relation.
    D. ordinate.
    E. function.

23. We call the dependent-variable coordinate (usually *y*) of a point on the Cartesian plane the
    A. magnitude.
    B. abscissa.
    C. relation.
    D. ordinate.
    E. function.

24. We can use an uncalibrated straight edge and a pencil alone to
    A. construct a line segment passing through a single defined point.
    B. construct a line segment connecting two defined points.
    C. construct a triangle connecting three defined points.
    D. construct a quadrilateral connecting four defined points.
    E. All of the above

25. A half-open line segment
    A. extends infinitely far in one direction.
    B. contains neither of its end points.
    C. extends infinitely far in both directions.
    D. contains both of its end points.
    E. contains one of its end points but not the other.

26. In order for a plane quadrilateral to constitute a trapezoid, one pair of opposite sides must be parallel and no sides may meet except at their end points. What other requirement, if any, must a quadrilateral fulfill in order to "qualify" as a trapezoid?
    A. None!
    B. All four angles must have the same measure.
    C. All four sides must have the same length.
    D. Both diagonals must have the same length.
    E. The diagonals must intersect at a right angle.

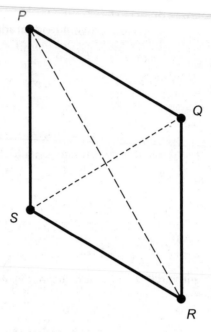

FIGURE TEST I-4 • Illustration for Part I
Test Question 27.

27. Figure Test I-4 illustrates a geometric figure that lies entirely in a single plane, and all four of whose sides measure the same length. Based on this knowledge, we can be absolutely certain that

   A. all four of the triangles formed by the outer sides and the half-diagonals are directly congruent.
   B. all four of the triangles formed by the outer sides and the half-diagonals are directly similar.
   C. the two diagonals intersect at a right angle.
   D. the sum of the lengths of the diagonals equals the perimeter of the whole figure.
   E. All of the above

28. Consider two distinct lines *L* and *M* that lie in the same plane. Suppose that both *L* and *M* intersect a third line *N*, and both *L* and *M* run perpendicular to *N*. In this situation, we can have total confidence that *L* and *M* constitute

   A. skew lines.
   B. perpendicular lines.
   C. complementary lines.
   D. parallel lines.
   E. congruent lines.

29. Imagine two triangles, both of which have one side measuring 10 units in length. In both triangles, the interior angle at either end of the 10-unit side measures 40°. Based on this information, we know that these triangles

   A. exhibit direct similarity.
   B. have equal interior areas.
   C. have equal perimeters.
   D. All of the above
   E. None of the above

30. Two triangles are directly congruent if and only if they're directly similar and

   A. corresponding angles have equal measures, going around both triangles counterclockwise.
   B. corresponding angles have equal measures, going around one triangle clockwise and the other triangle counterclockwise.
   C. corresponding sides have the same lengths, going around both triangles counterclockwise.
   D. corresponding sides have the same lengths, going around one triangle clockwise and the other triangle counterclockwise.
   E. Any of the above

31. Imagine a triangle with interior angles measuring 10°, 20°, and 150°. From this information, we can have complete confidence that the figure constitutes

   A. a right triangle.
   B. a non-Euclidean triangle.
   C. an isosceles triangle.
   D. an equilateral triangle.
   E. an obtuse triangle.

32. Imagine two lines L and M that intersect at a point P. In this situation, any pair of adjacent angles between L and M is

   A. congruent.
   B. acute.
   C. obtuse.
   D. supplementary.
   E. transverse.

33. Imagine a triangle with interior angles measuring $\pi/6$ rad, $\pi/3$ rad, and $\pi/2$ rad. From this information, we can have complete confidence that the figure constitutes

   A. a right triangle.
   B. a non-Euclidean triangle.
   C. an isosceles triangle.
   D. an equilateral triangle.
   E. an obtuse triangle.

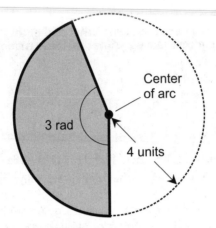

FIGURE TEST I-5 • Illustration for Part I
Test Questions 34 and 35.

34. **What's the interior area of the shaded region in Fig. Test I-5? Assume that the entire curve (including the dashed portion) is a perfect circle, and that the center of the arc lies at the center of the circle.**

    A. 4π square units
    B. 6π square units
    C. 18 square units
    D. 24 square units
    E. We need more information to calculate it.

35. **What's the perimeter of the shaded region in Fig. Test I-5? Assume that the entire curve (including the dashed portion) is a perfect circle, and that the center of the arc lies at the center of the circle.**

    A. 3π units
    B. 4π units
    C. 11 units
    D. 20 units
    E. We need more information to calculate it.

36. **Which of the following statements accurately expresses the parallel principle as it applies to Euclidean geometry in a single plane?**

    A. Suppose that *L* represents a line and *P* represents a point that doesn't lie on *L*. There exist no lines through *P* that run parallel to *L*.
    B. Suppose that *L* represents a line and *P* represents a point that doesn't lie on *L*. There exist two lines *M* and *N* through *P*, such that *M* and *N* both run parallel to *L*.
    C. Suppose that *L* represents a line and *P* represents a point that doesn't lie on *L*. There exist infinitely many lines through *P* that run parallel to *L*.
    D. Suppose that *L* represents a line and *P* represents a point that doesn't lie on *L*. The number of lines through *P* that run parallel to *L* depends on the distance between *P* and *L*.
    E. Suppose that *L* represents a line and *P* represents a point that doesn't lie on *L*. There exists one and only one line *M* through *P*, such that *M* runs parallel to *L*.

37. In order for a plane quadrilateral to constitute a rhombus, any two opposite sides must run parallel to each other. What other requirement, if any, must a quadrilateral fulfill in order to "qualify" as a rhombus?

    A. None!
    B. All four angles must have the same measure.
    C. All four sides must have the same length.
    D. Both diagonals must have the same length.
    E. The figure must have the same interior area as a square of the same perimeter.

38. Imagine two triangles, both of which have interior angles measuring 50°, 60°, and 70° in that order as we proceed around them counterclockwise. Based on this information, we can have complete confidence that the two triangles

    A. exhibit direct similarity.
    B. have equal interior areas.
    C. have equal perimeters.
    D. All of the above
    E. None of the above

39. Consider the following equation that represents a straight line in Cartesian coordinates:

$$-4x + y = 5$$

    What's the slope of this line? (Here's a hint: Use a little algebra to get the equation into the slope-intercept form.)

    A. 4
    B. −5
    C. −5/4
    D. 4/5
    E. We need more information to calculate it.

40. Imagine a perfectly square, flat field surrounded by four straight lengths of fence. You build a straight fence diagonally across the field, dividing the field into two triangles, both of which are

    A. right triangles.
    B. isosceles triangles.
    C. directly congruent.
    D. directly similar.
    E. All of the above

41. How far from the origin does point P lie in Fig. Test I-6?

    A. The square roof of 12 units
    B. The square root of 14 units
    C. 7/2 units
    D. 5 units
    E. We need more information to calculate this distance.

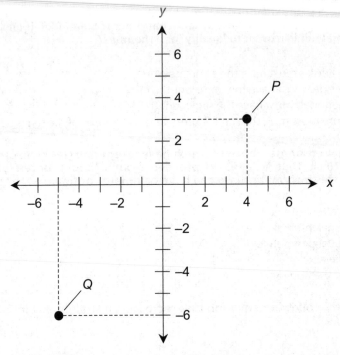

FIGURE TEST I-6 • Illustration for Part I Test Questions 41 through 43.

42. **How far from the origin does point Q lie in Fig. Test I-6?**
    A. The square root of 61 units.
    B. 9 units
    C. 8 units
    D. 11/2 units
    E. We need more information to calculate this distance.

43. **What's the distance between points P and Q in Fig. Test I-6?**
    A. 14 units
    B. The square root of 162 units
    C. 5 plus the square root of 61 units
    D. 13 units
    E. We need more information to calculate this distance.

44. **Consider a plane polygon having n sides. Let θ_1, θ_2, θ_3, ..., θ_n represent the interior angles. If we express the angular measures in radians, then**

$$\theta_1 + \theta_2 + \theta_3 + ... + \theta_n = \pi (n - 2)$$

Based on this formula and on our knowledge of the relation between degrees and radians, what's the sum of the measures of the interior angles of a 20-sided plane polygon in degrees?

A. 1620°
B. 1800°
C. 3240°
D. 3600°
E. 6480°

45. Suppose you draw a line L and a point P near that line. Then you drop a perpendicular from point P to line L, and let Q represent the point where the perpendicular intersects L. Then you draw a point R on line L, different from point Q. You can have complete confidence that the points P, Q, and R lie at the vertices of

A. an equilateral triangle.
B. a similar triangle.
C. an isosceles triangle.
D. a right triangle.
E. an obtuse triangle.

46. In order for a plane quadrilateral to constitute a parallelogram, any two opposite sides must run parallel to each other and no two sides may meet except at their end points. What other requirement, if any, must a quadrilateral fulfill in order to "qualify" as a parallelogram?

A. None!
B. All four angles must have the same measure.
C. All four sides must have the same length.
D. Both diagonals must have the same length.
E. The figure must have the same interior area as a rectangle of the same perimeter.

47. If we consider the rotational sense important when we express an angle θ, then *clockwise* angular motion means that

A. $\theta = 0$ rad.
B. $\theta < 0$ rad.
C. $\theta > 0$ rad.
D. $-\pi$ rad $< \theta < \pi$ rad.
E. -2π rad $< \theta < 2\pi$ rad.

48. A closed-ended ray

A. extends infinitely far in one direction.
B. contains neither of its end points.
C. extends infinitely far in both directions.
D. contains both of its end points.
E. has finite length.

49. Imagine two triangles, both of which have equal perimeters. Based on this information, we know for certain that the two triangles
    A. exhibit direct similarity.
    B. have equal interior areas.
    C. have corresponding interior angles of equal measure.
    D. All of the above
    E. None of the above

50. Imagine that we *circumscribe* a circle C with a regular polygon P_c having n sides (where n represents a positive integer larger than 3), and then we increase n without limit. Also suppose that we *inscribe* the same circle with another regular polygon P_i having the same number of sides as P_c at all times. As we make n grow larger indefinitely, all the while ensuring that P_c and P_i both fit "snugly" against C, the interior areas of P_c and P_i both approach
    A. $\pi^2/10$ times the interior area of C.
    B. $\pi/4$ times the interior area of C.
    C. $\pi/3$ times the interior area of C.
    D. $\pi/2$ times the interior area of C.
    E. the interior area of C.

Part II

Three Dimensions and Up

An Expanded Set of Rules

In *solid geometry*, we have an extra dimension compared with plane geometry. We have greater freedom, but with that freedom comes complexity, reflecting the expanded range of maneuvers that we must learn and master.

CHAPTER OBJECTIVES

In this chapter, you will

- Define elementary objects in three dimensions.
- Learn how elementary objects interact in three dimensions.
- Discover how angles and distances relate in three dimensions.
- Learn the fundamental principles of solid geometry.
- Explore the behavior of parallel and intersecting planes and lines.

Points, Lines, Planes, and Space

We can imagine a point in space as an infinitely tiny ball having height, width, and depth all equal to zero, but nevertheless possessing a specific location. A point is *zero-dimensional* (0D). A point in space therefore constitutes the same sort of object as does a point in a plane or a point on a line.

We can imagine a line in space as an infinitely thin, perfectly straight, infinitely long wire—the same sort of object as a line in two dimensions. A straight line is *one-dimensional* (1D). Although lines in space are just like lines in planes, a line in space can run in more different directions than a line confined to a single plane.

We can imagine a *plane* as an infinitely thin, perfectly flat surface having an infinite expanse, like an unlimited, flat sheet of paper thinner than anything that could ever exist in the real world. A plane is *two-dimensional* (2D); in effect it's a "flat 2D universe" in which all the rules of Euclidean plane geometry apply.

Space comprises the set of points for all possible physical locations in the universe as we perceive it. Space is *three-dimensional* (3D). We ignore time, often called a "fourth dimension," when we work in *Euclidean 3D space*. However, we can define an alternative form of 3D space (or *three-space*) having two spatial dimensions and one time dimension. We might imagine this type of three-space as a Euclidean plane in which we account for time past, present, and future.

If we allow for the passage of time, or perhaps even free time travel, along with Euclidean three-space, we get *four-dimensional* (4D) space, also known as *four-space* or *hyperspace*. We'll take a look at some properties of hyperspace later in this course. As you can imagine, hyperspace gives us "hyperfreedom"—and "hypercomplexity" as well.

Naming Points, Lines, and Planes

Points, lines, and planes in solid geometry usually bear names consisting of uppercase, italicized letters of the alphabet, just as they do in plane geometry. We'll commonly name a point *P*, *Q*, or *R*, and a line *L*, *M*, or *N*. When we want to name planes in 3D space, the letters *X*, *Y*, and *Z* make good choices.

If we encounter a situation involving a lot of points, lines, and/or planes, we can use a single letter for each type of object and attach numeric subscripts.

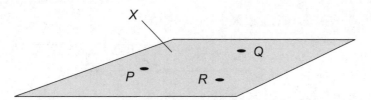

FIGURE 7-1 · Three points *P*, *Q*, and *R*, not all on the same line, define a specific plane *X*. The plane extends infinitely in 2D.

Therefore, we might have points called P_1, P_2, P_3, P_4, P_5 and so forth, lines called L_1, L_2, L_3, L_4, L_5 and so forth, and planes called X_1, X_2, X_3, X_4, X_5 and so forth.

Three-Point Principle

Suppose that *P*, *Q* , and *R* represent three different geometric points, no two of which lie on the same line. These points define *one and only one* (i.e., a *unique* or *specific*) plane *X*. The following two statements always hold true, as shown in Fig. 7-1:

- *P*, *Q*, and *R* lie in a single plane *X*.
- *X* constitutes the only plane in which all three points lie.

We always need at least three points to uniquely define a plane in Euclidean three-space. It's possible, however, that more than three points—even infinitely many—can all lie in the same plane.

Still Struggling

In order to diagram the fact that a surface extends infinitely in 2D, we must use our imaginations. Our task is more difficult than showing that a line extends infinitely in 1D, because we can't conveniently draw arrows on the edges of a plane region the way we can draw them on the ends of a line segment. Geometers and draftspeople sometimes draw planes as rectangles in perspective, so that they appear as parallelograms or trapezoids when rendered on a flat page. However, when we draw a plane in a diagram, we should always make sure that our readers know we don't intend to show a quadrilateral of finite extent rather than a plane of infinite extent!

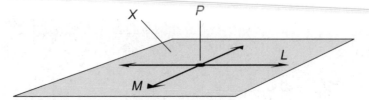

FIGURE 7-2 • Two lines *L* and *M*, intersecting at point *P*, define a specific plane *X*. The plane extends infinitely in 2D.

Intersecting Line Principle

Suppose that two distinct lines L and M intersect in a point P. In that case, the two lines together define a unique plane X. The following statements always hold true, as shown in Fig. 7-2:

- L and M lie in a single plane X.
- X constitutes the only plane in which both lines lie.

We always need at least two intersecting lines to uniquely define a plane in Euclidean three-space. It's possible, however, that more than two intersecting lines—even infinitely many—can all lie in the same plane.

Line and Point Principle

Let L represent a line, and let P represent a point that doesn't lie on L. In this situation, line L and point P define a unique plane X. The following two statements always hold true:

- L and P lie in a single plane X.
- X constitutes the only plane in which both L and P lie.

Plane Regions

The 2D counterpart of the 1D line segment is a "piece of a plane" called a *simple plane region*. A simple plane region consists of all the points inside a polygon or enclosed curve. The points that we consider to fall inside a simple plane region might include all, some, or none of the points that lie on the enclosing polygon or curve itself.

- If the region includes all of the points on the enclosing figure, we call the region *closed*.

- If the region includes some but not all of the points on the enclosing figure, we call the region *partially closed* or *partially open*.
- If the region includes none of the points on the enclosing figure, we call the region *open*.

Figure 7-3 shows examples of the above-described types of regions. At A, we see closed regions; at B, we see partially open regions; at C we see open regions.

- When we want to include part or all of the boundary, we draw the included portion as a solid line.
- When we want to exclude part or all of the boundary, we draw the excluded part as a dashed line.
- When we want to include a particular boundary point, we draw it as a solid black dot.
- When we want to exclude a particular boundary point, we draw it as a small open circle.

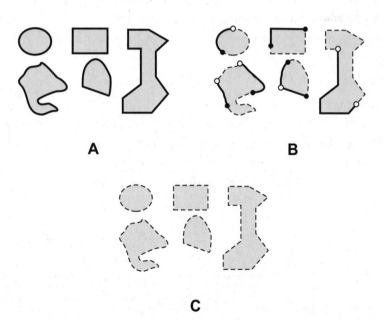

FIGURE 7-3 • At A, closed plane regions. At B, partially open plane regions. At C, open plane regions. Black dots and solid lines indicate included boundary points; small open circles and dashed lines denote nonincluded boundary points.

TIP   *The corresponding regions in Figs. 7-3A, B, and C have identical shapes. They also have identical perimeters and identical interior areas. The inclusion of part (or all) of the outer boundary adds no perimeter or interior area to the region. The lack of part (or all) of the outer boundary takes away nothing from the perimeter or interior area of the region.*

Still Struggling

The examples in Fig. 7-3 show specialized scenarios in which the plane regions are contiguous, or "all of a piece." Some plane regions consist of two or more noncontiguous subregions. If you work in mathematics long enough, you'll eventually encounter a plane region with characteristics so complicated that you'll have trouble figuring out how to define it, let alone manipulate it. You need not concern yourself with such things here, other than to acknowledge their existence. If you plan to become a serious student of geometry, and especially if you want to become a mathematics teacher or professor, you should know that somewhere in the vast expanse of Euclidean space, these beasts await you. When you find them, you'll have great fun!

Half Planes

Mathematicians occasionally talk about the portion of a geometric plane that lies "on one side" of a certain line. Look at Fig. 7-4 and imagine the *union* (the geometric combination) of all possible rays that start at L, then pass through line M (which runs parallel to L), and extend onward past M forever in one direction. The region thus defined constitutes a *half plane*.

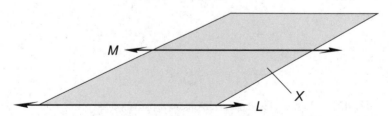

FIGURE 7-4 • A half plane X, defined by two parallel lines, L and M. The half plane extends infinitely in 2D on the "M" side of L.

The half plane defined by L and M might include the end line L, in which case we call it *closed-ended*. Then we draw line L as a solid line, as it appears in Fig. 7-4. But the end line might not be part of the half plane, in which case the half plane is *open-ended*. In that case we draw L as a dashed line.

Parts of the end line might lie in the half plane while other parts don't. Infinitely many situations of this kind exist! We can illustrate relatively simple cases by making some parts of L solid and other parts dashed, all the while remembering to use solid black dots to represent included points and small open circles to represent nonincluded points.

Intersecting Planes

Suppose that two different planes X and Y have some points in common. In this type of situation, we'll always find that the two planes intersect in a unique straight line L. The following statements always hold true, as shown in Fig. 7-5:

- Planes X and Y share a single line L.

- L constitutes the only line that lies in both planes X and Y.

Parallel Lines in 3D Space

By definition, two different lines L and M in three-space are *parallel lines* if and only if both of the following statements hold true:

- Lines L and M do not intersect at any point.

- Lines L and M lie in the same plane X.

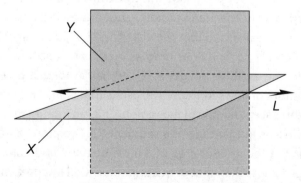

FIGURE 7-5 • The intersection of two planes X and Y determines a unique line L. The planes extend infinitely in 2D.

If two lines run parallel to each other and if they both lie in a given plane X, then X constitutes the only plane in which the two lines lie. Therefore, we can say that two parallel lines define a unique plane in Euclidean three-space.

Skew Lines

By definition, two lines L and M in three-space constitute *skew lines* (or they *run askew*) and only if both of the following statements hold true:

- Lines L and M do not intersect at any point.
- Lines L and M do not lie in the same plane (so they don't run parallel to each other).

TIP *Imagine an infinitely long, straight two-lane highway and an infinitely long, straight electrical cable propped up on utility poles. Further imagine that the electrical cable and the highway centerline are both infinitely thin, and that the electrical cable doesn't sag between the poles. Suppose that the electrical cable passes over the highway somewhere, but does not run parallel to the highway. In that case, the highway centerline and the electrical cable define skew lines.*

PROBLEM 7-1

Find an example of a theoretical plane region with a finite, nonzero area but an infinite perimeter.

SOLUTION

Examine Fig. 7-6. Suppose that the three lines *PQ*, *RS*, and *TU* (none of which form part of the plane region *X*, but are shown only for reference) run mutually parallel, and that the distances $d_1, d_2, d_3, \ldots$ are such that d_2 is half as long as d_1, d_3 is half as long as d_2, d_4 is half as long as d_3, and, in general, for any positive integer n, $d_{(n+1)}$ is half as long as d_n. Also suppose that the length of line segment *PV* exceeds that of line segment *PT*. In this rather bizarre scenario, the plane region *X* has an infinite number of sides, each of which is longer than line segment *PT*. Therefore, *X* has an infinite perimeter. But the interior area of *X* must be finite and nonzero, because the area of *X* is less than that of quadrilateral *PQSR* but greater than that of quadrilateral *TUSR*.

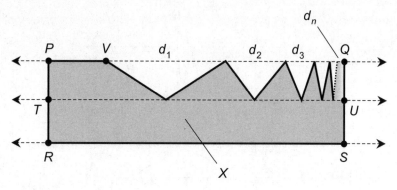

FIGURE 7-6 · Illustration for Problem 7-1.

PROBLEM **7-2**

How many planes can mutually intersect in a single straight line?

SOLUTION

In theory, an infinite number of planes can all intersect along a single line. Think of the line as a "Euclidean hinge," and then imagine a plane that can swing freely around the hinge. Each position of the "swinging plane" represents a unique plane in space.

Angles and Distances

Let's define the *angles between intersecting planes*, and then explore how these angles behave. Let's do the same with the *angles between an intersecting line and plane*.

Angles between Intersecting Planes

In Fig. 7-7, two planes X and Y intersect along a specific line L. Consider line M in plane X and line N in plane Y, such that M runs perpendicular to L (a fact that we can write in "shorthand" as $M \perp L$) and N also runs perpendicular to L ($N \perp L$). Lines M and N don't necessarily run perpendicular to each other, although they might. In a case of this sort, we call the angle between the intersecting planes X and Y a *dihedral angle*. Its measure equals the measure of the angle between lines M and N. The jargon "dihedral" means "two-faced."

We can represent a dihedral angle between two intersecting planes X and Y in two ways when we look at Fig. 7-7. We might speak of the smaller (acute or

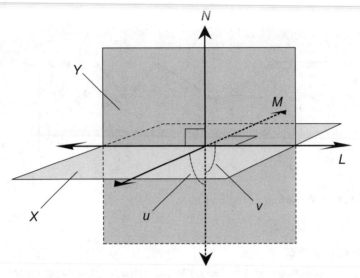

FIGURE 7-7 · Two intersecting planes, each containing a line. See text for discussion.

right) angle between lines M and N, whose measure we denote by u. Alternatively, we might consider the larger (obtuse or right) angle between lines M and N, whose measure we denote by v.

TIP *If you see only one dihedral angle mentioned when you encounter two intersecting planes, the author usually wants you to think of the smaller of the two possible angles. Therefore, in most situations, the measure of a dihedral angle is always positive, but it never exceeds a right angle ($0° < u \leq 90°$ or $0 < u \leq \pi/2$).*

Adjacent Dihedral Angles

Suppose that two planes intersect, and we call their angles of intersection u and v as defined earlier. If we specify the measures of u and v in degrees, then

$$u + v = 180°$$

and if we specify the measures of u and v in radians, then

$$u + v = \pi$$

Perpendicular Planes

Suppose that two planes X and Y intersect along a single line L. Consider line M in plane X and line N in plane Y, such that $M \perp L$ and $N \perp L$, as shown in Fig. 7-7. We say that X and Y constitute *perpendicular planes* if and only if the

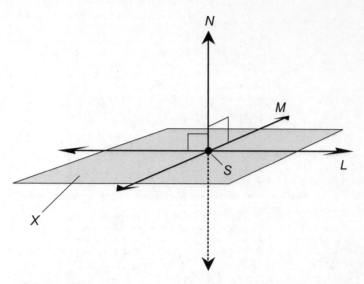

FIGURE 7-8 · Line *N* through plane *X* at point *S* runs normal to *X* and only if *N* runs perpendicular to some line *L* and *N* also runs perpendicular to another line *M*, where *L* and *M* both lie in plane *X* and intersect at point *S*.

angles between lines *M* and *N* are right angles, that is, if and only if $u = v = 90°$ ($\pi/2$). Actually, it suffices to say that either $u = 90°$ ($\pi/2$) or $v = 90°$ ($\pi/2$).

Normal Line to a Plane

Look at Fig. 7-8, and imagine that we can uniquely define a plane *X* on the basis of two lines *L* and *M* that intersect each other at a single point *S*. In this type of situation, the line *N* that passes through plane *X* at point *S* runs *normal* (also called *perpendicular* or *orthogonal*) to plane *X* if and only if $N \perp L$ and $N \perp M$. Line *N* is the only line normal to plane *X* at point *S*. Furthermore, line *N* runs perpendicular to any line, line segment, or ray that lies in plane *X* and passes through point *S*.

Angle between an Intersecting Line and Plane

Let *X* represent a plane as shown in Fig. 7-9. Suppose that a line *O*, which does not necessarily run normal to plane *X*, intersects *X* at a point *S*. In order to define an angle at which line *O* intersects plane *X*, let's construct three "scaffolding" objects, as follows:

- Let *N* represent a line normal to plane *X*, passing through point *S*
- Let *Y* represent the plane determined by the intersecting lines *N* and *O*
- Let *L* represent the line formed by the intersection of planes *X* and *Y*

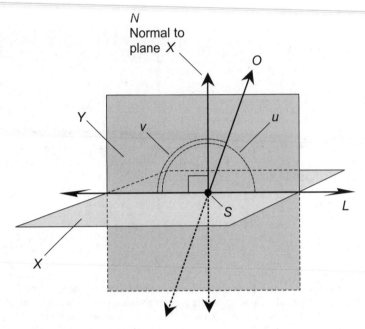

FIGURE 7-9 · Angles u and v between a plane X and a line O that passes through X at point S.

We can describe the angle between line O and plane X in two ways. The first angle, whose measure we denote as u, is the smaller (acute or right) angle between lines L and O as determined in plane Y. The second angle, whose measure we denote as v, is the larger (obtuse or right) angle between lines L and O as determined in plane Y.

TIP *If only one angle is mentioned, then we should consider the "angle between a line and a plane that intersect" as the smaller angle u. Therefore, the angle of intersection is positive but never larger than a right angle ($0° < u \leq 90°$ or $0 < u \leq \pi/2$).*

Adjacent Line/Plane Angles

Suppose that a line and a plane intersect, and we call their angles of intersection u and v as defined earlier and as shown in Fig. 7-9. If we specify u and v in degrees, then

$$u + v = 180°$$

and if we specify u and v in radians, then

$$u + v = \pi$$

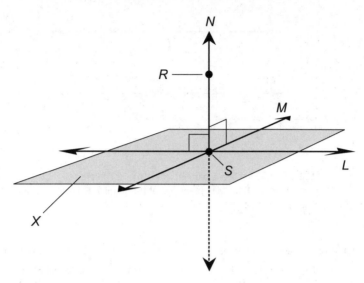

FIGURE 7-10 · Line *N* through point *R* is normal to plane *X* at point *S*. The distance between *R* and *X* equals the length of line segment *RS*.

Dropping a Normal to a Plane

Let *R* represent a point near, but not in, some known plane *X*. In that case, there exists exactly one line *N* through point *R*, intersecting plane *X* at some point *S*, such that line *N* runs normal to plane *X* as shown in Fig. 7-10. Any line within plane *X* that passes through point *S*, such as *L* or *M* shown in the figure, must run perpendicular to line *N*.

Distance between a Point and Plane

Suppose that *R* represents a point near, but not in, a plane *X*. Let *N* represent the unique line through *R* that runs normal to plane *X*. Suppose that line *N* intersects plane *X* at point *S*. We define the distance between point *R* and plane *X* as the length of line segment *RS* as shown in Fig. 7-10.

TIP *Whenever we talk or write about "the distance between a point and a plane," we mean to specify the shortest possible distance (as shown in Fig. 7-10) unless we explicitly define it as something else.*

Plane Perpendicular to Line

Imagine a line *N* in space. Imagine a specific point *S* on line *N*. There exists exactly one plane *X* containing point *S*, such that line *N* runs normal to plane

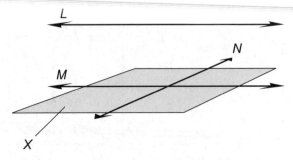

FIGURE 7-11 · A line *L* parallel to a plane *X*. There exist infinitely many lines *M* in plane *X* that run parallel to *L*; all other lines *N* in plane *X* are skew lines relative to *L*.

X at point *S* (Fig. 7-10). As before, any line in plane *X* that passes through point *S*, such as *L* or *M* shown in the figure, must run perpendicular to line *N*.

Line Parallel to Plane

We say that a line *L* runs parallel to a plane *X* in Euclidean three-space if and only if *L* shares no points in common with *X*. In a situation like this, we can find infinitely many lines *M* in plane *X*, such that *L* and *M* constitute parallel lines as shown in Fig. 7-11. Any line *N* in plane *X*, other than line *M*, constitutes a skew line relative to *L*.

Distance between Parallel Line and Plane

Suppose that a given line *L* runs parallel to a given plane *X*. Let *R* represent a point on line *L*. We define the distance between line *L* and plane *X* as the distance between point *R* and plane *X*.

TIP *Whenever we talk or write about "the distance between a line and a plane parallel to that line," we mean to specify the shortest possible distance unless we explicitly define it as something else.*

Addition and Subtraction of Angles between Intersecting Planes

Angles between intersecting planes add and subtract in the same fashion as angles between intersecting lines (or line segments) do. We can prove this fact, based on knowledge that we already have.

Suppose that three planes *X*, *Y*, and *Z* intersect in a single, common line *L*, as shown in Fig. 7-12. Let *S* represent a point on line *L*. Let *P*, *Q*, and *R* represent points on planes *X*, *Y*, and *Z*, respectively, such that each of the three line

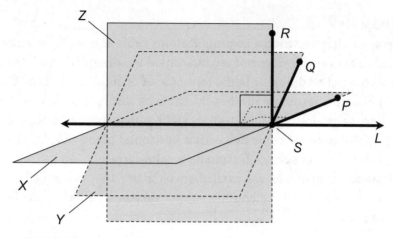

FIGURE 7-12 · Addition and subtraction of angles between planes.

segments *SP*, *SQ*, and *SR* runs perpendicular to line *L*. Let $\angle XY$ represent the angle between planes *X* and *Y*, $\angle YZ$ represent the angle between planes *Y* and *Z*, and $\angle XZ$ represent the angle between planes *X* and *Z*. From the preceding definition of the angle between two planes, we know the following three facts:

$$\angle XY = \angle PSQ$$

$$\angle YZ = \angle QSR$$

$$\angle XZ = \angle PSR$$

We know that line segments *SP*, *SQ*, and *SR* all lie in a single plane, because they all intersect at point *S* and they all run perpendicular to line *L*. From the rules for addition of angles in a plane, we also know that the following three statements hold true for the measures of the angles between the line segments:

$$\angle PSQ + \angle QSR = \angle PSR$$

$$\angle PSR - \angle QSR = \angle PSQ$$

$$\angle PSR - \angle PSQ = \angle QSR$$

Substituting the angles between the planes for the angles between the line segments, we see that the following three statements all hold true for the measures of the angles between the planes:

$$\angle XY + \angle YZ = \angle XZ$$

$$\angle XZ - \angle YZ = \angle XY$$

$$\angle XZ - \angle XY = \angle YZ$$

PROBLEM 7-3

Imagine that we string a communications cable above a freshwater lake, such that the cable does not sag but runs along a perfectly horizontal line. We've attached the cable to the tops of a set of utility poles. The engineering literature recommends that the cable be suspended 10 meters above "effective ground." The literature also tells us that, for a body of freshwater, "effective ground" coincides with a horizontal plane that lies 2 meters below the water surface (assuming a calm surface). How tall should the poles be? Assume that we install them all so that they're perfectly vertical, and that they're all tall enough so that we can set them securely in the lake bottom.

SOLUTION

Because the poles are perfectly vertical, they stand perpendicular to the surface of the lake. Therefore, the pole tops should all be 10 meters above "effective ground." It follows that the poles should each extend (10 − 2) meters, or 8 meters, above the water surface. The overall height of each pole will depend on the depth of the lake at the point where we place it, and on the depth into the bottom to which we must set it to ensure that it remains standing upright.

PROBLEM 7-4

You fly a kite over a perfectly flat, horizontal field. The design of the kite causes it to fly at a "high angle," meaning that the kite string runs nearly straight up and down. Suppose that the kite line does not sag, and the kite flies at an angle 10° away from the vertical. Imagine that the sun shines down from exactly the zenith (straight overhead). What's the angle between the kite string and its shadow on the field?

SOLUTION

Let's say that you stand at a point called S on the surface of the field called plane X, as shown in Fig. 7-13. The kite line and its shadow lie along lines SR and ST. (Point T does not necessarily represent the shadow of the kite, however.) The sun shines down so that its rays run along and parallel to line SQ, which runs normal to plane X and passes through point S. Lines SQ, SR, and ST all lie in a common plane Y, which is oriented perpendicular to plane X. You know that the measure of $\angle RSQ$ equals 10°, because you've been

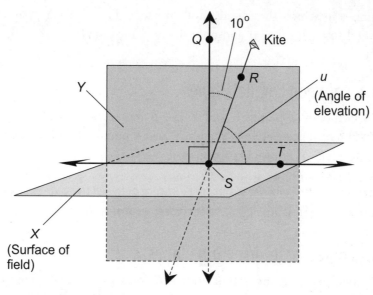

FIGURE 7-13 · Illustration for Problem 7-4.

given this information. You also know that the measure of $\angle TSQ$ equals 90°, because line QS runs normal to plane X, and line ST lies in plane X. Because lines SQ, SR, and ST all lie in the same plane Y, you can conclude that

$$\angle TSR + \angle RSQ = \angle TSQ$$

and therefore that

$$\angle TSR = \angle TSQ - \angle RSQ$$

The measure of $\angle TSR$, which represents the angle between the kite line and its shadow, equals 90° – 10°, or 80°.

More Facts

Lines, planes, and angles behave according to specific principles in Euclidean three-space. Let's briefly examine a few of these rules.

Parallel Planes

Two distinct planes run parallel to each other in three-space if and only if they do not intersect. Two distinct half planes run parallel to each other if and only if the planes in which they lie do not intersect. Two distinct plane regions run parallel to each other if and only if the planes in which they lie do not intersect.

Distance between Parallel Planes

Consider two parallel planes X and Y. Let R represent an arbitrary point on plane X. The distance between planes X and Y equals the distance between point R and plane Y, as previously defined.

TIP *Whenever we talk or write about "the distance between two planes," we mean to specify the shortest possible distance unless we state otherwise.*

Vertical Angles for Intersecting Planes

Consider two planes Y and Z that intersect along a line L. Also consider five points P, Q, R, S, and T as shown in Fig. 7-14, such that all of the following conditions hold true:

- Point T lies at the intersection of lines L, PS, and QR.
- Points Q and R lie in plane Y.
- Points P and S lie in plane Z.
- Lines PS and QR both run perpendicular to line L.

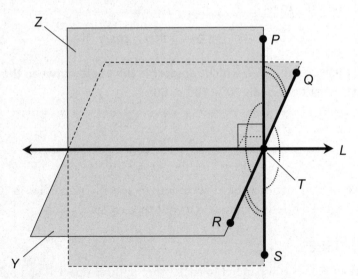

FIGURE 7-14 · Vertical angles between intersecting planes.

In this situation, $\angle QTP$ and $\angle RTS$ are vertical angles. Also, $\angle PTR$ and $\angle STQ$ are vertical angles. Therefore, $\angle QTP$ has the same measure as $\angle RTS$, and $\angle PTR$ has the same measure as $\angle STQ$.

Alternate Interior Angles for Intersecting Planes

Consider a plane X that passes through two parallel planes Y and Z, intersecting Y and Z in lines L and M, respectively. Define points P, Q, R, S, T, U, V, and W as shown in Figs. 7-15A and Fig. 7-15B, such that all of the following conditions hold true:

- Point V lies at the intersection of lines L, PQ, and RS.
- Point W lies at the intersection of lines M, PQ, and TU.
- Points P and Q lie in plane X.
- Points R and S lie in plane Y.
- Points T and U lie in plane Z.
- Lines PQ and RS both run perpendicular to line L.
- Lines PQ and TU both run perpendicular to line M.

In this scenario, $\angle RVP$ and $\angle UWQ$ are alternate interior angles (Fig. 7-15A). Also, $\angle QWT$ and $\angle PVS$ are alternate interior angles (Fig. 7-15B). Alternate interior angles always have equal measures. Therefore, $\angle RVP$ has the same measure as $\angle UWQ$, and $\angle QWT$ has the same measure as $\angle PVS$.

Alternate Exterior Angles for Intersecting Planes

Let X represent a plane that passes through two parallel planes Y and Z, intersecting Y and Z in lines L and M, respectively. Define points P, Q, R, S, T, U, V, and W as shown in Fig. 7-16, such that all of the following conditions hold true:

- Point V lies at the intersection of lines L, PQ, and RS.
- Point W lies at the intersection of lines M, PQ, and TU.
- Points P and Q lie in plane X.
- Points R and S lie in plane Y.
- Points T and U lie in plane Z.
- Lines PQ and RS both run perpendicular to line L.
- Lines PQ and TU both run perpendicular to line M.

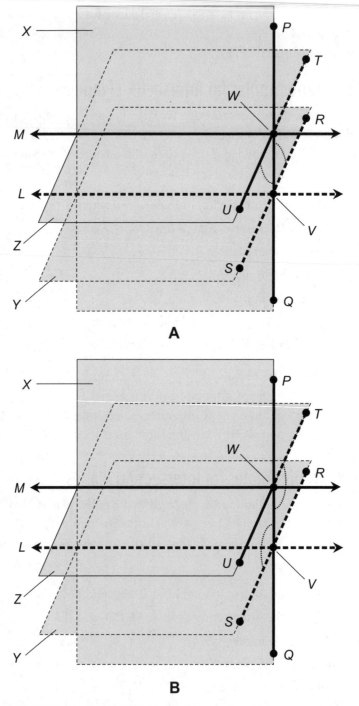

FIGURE 7-15 · A. Alternate interior angles between intersecting planes.
B. Another example of alternate interior angles between intersecting planes.

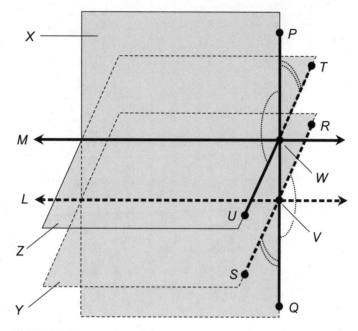

FIGURE 7-16 · Alternate exterior angles between intersecting planes.

In this case, ∠*TWP* and ∠*SVQ* are alternate exterior angles. Also, ∠*PWU* and ∠*QVR* are alternate exterior angles. Alternate exterior angles always have equal measures. Therefore, ∠*TWP* has the same measure as ∠*SVQ*, and ∠*PWU* has the same measure as ∠*QVR*.

Corresponding Angles for Intersecting Planes

Let *X* represent a plane that passes through two parallel planes *Y* and *Z*, intersecting *Y* and *Z* in lines *L* and *M*, respectively. Define points *P*, *Q*, *R*, *S*, *T*, *U*, *V*, and *W* as shown in Fig. 7-17, such that all of the following conditions hold true:

- Point *V* lies at the intersection of lines *L*, *PQ*, and *RS*.
- Point *W* lies at the intersection of lines *M*, *PQ*, and *TU*.
- Points *P* and *Q* lie in plane *X*.
- Points *R* and *S* lie in plane *Y*.
- Points *T* and *U* lie in plane *Z*.
- Lines *PQ* and *RS* both run perpendicular to line *L*.
- Lines *PQ* and *TU* both run perpendicular to line *M*.

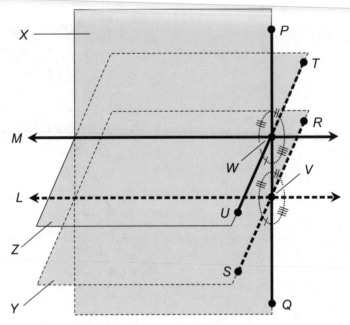

FIGURE 7-17 · Corresponding angles between intersecting planes.

In this case, the following equations describe pairs of corresponding angles, where each pair has equal measures:

$$\angle UWQ = \angle SVQ$$
$$\angle TWP = \angle RVP$$
$$\angle PWU = \angle PVS$$
$$\angle QWT = \angle QVR$$

Parallel Principle for Planes

Consider a plane X along with some point R that does not lie on X. In Euclidean three-space, there exists one and only one plane Y through R such that plane Y runs parallel to plane X. This statement expresses the 3D counterpart of the parallel principle for 2D Euclidean geometry. We can deny the 3D parallel principle and nevertheless have a workable mathematical system, just as we can deny the 2D parallel principle. When we deny the parallel principle in Euclidean three-space, we obtain one or the other of the following situations:

- There can exist *more than one*, and perhaps *infinitely many*, planes Y through point R such that plane Y runs parallel to plane X.
- There can exist *no* plane Y through point R such that plane Y runs parallel to plane X.

TIP *Either of the foregoing hypotheses gives rise to a form of non-Euclidean geom-*
etry in which three-space has "curvature" like the 2D surface of a funnel, cylinder,
or sphere, but with an added dimension. The German mathematicians Karl
Friedrich Gauss *(1777-1855) and* Bernhard Riemann *(1826-1866) developed*
detailed theories in non-Euclidean geometry. Later, Albert Einstein (1879-1955)
dared to envision a non-Euclidean three-space universe in a literal sense. We'll
explore non-Euclidean geometry in Chap. 11.

Parallel Principle for Lines and Planes

Once again, consider a plane X along with some point R that does not lie on X.
There exist an infinite number of lines through R that run parallel to plane X.
All of these lines lie in the plane Y through R such that plane Y runs parallel to
plane X.

Still Struggling

The denial of the parallel principle for planes, defined in the previous paragraph,
can result in the existence of no lines through R that run parallel to plane X. In
certain specialized instances, it can even result in the existence of exactly one
line through R that runs parallel to plane X. If you have trouble imagining scenarios such as these, don't worry. You must think in 4D—a mental trick that few
humans can perform, even when they've "armed" their minds with the power of
non-Euclidean mathematics.

PROBLEM 7-5

Imagine that you stand inside a large warehouse. The floor is flat and
horizontal. The ceiling is also flat and horizontal, everywhere at a uniform height of 5.455 meters above the floor. You have a flashlight with a
narrow beam. You hold the flashlight so that its bulb rests 1.025 meters
above the floor. You shine the beam at an angle upward toward the ceiling. The center of the beam strikes the ceiling 9.577 meters from the
point on the ceiling directly above the bulb. How long is the line segment representing the center of the light beam? Round your answer off
to two decimal places.

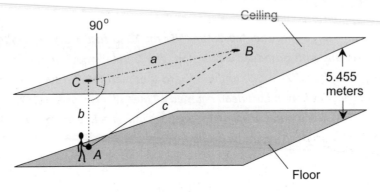

FIGURE 7-18 • Illustration for Problem 7-5.

✔ SOLUTION

Figure 7-18 shows a diagram for this situation. Call the flashlight bulb point A, the point at which the center of the light beam strikes the ceiling point B, and the point directly over the flashlight bulb point C. These three points define a triangle $\triangle ABC$. Now define the following three quantities for the sides of $\triangle ABC$:

- Let a represent the length of the side opposite point A.
- Let b represent the length of the side opposite point B.
- Let c represent the length of the side opposite point C.

In this situation, $\triangle ABC$ constitutes a right triangle, because line segment AC (whose length equals b) runs normal to the ceiling at point C, and therefore runs perpendicular to line segment BC (which lies on the ceiling). The right angle is $\angle ACB$. Based on this information, you know that the lengths of the sides of $\triangle ABC$ relate according to the Pythagorean equation

$$a^2 + b^2 = c^2$$

You want to know the length of side c. With the help of a little algebra, you can manipulate the above equation to obtain

$$c = (a^2 + b^2)^{1/2}$$

You've been told that side a measures 9.577 meters in length. The length of side b equals the height of the ceiling above the floor, minus the height of the bulb above the floor, so you can calculate that

$$b = 5.455 - 1.025$$
$$= 4.430 \text{ meters}$$

You can now solve for *c*, getting

$$c = (9.577^2 + 4.430^2)^{1/2}$$
$$= (91.719 + 19.625)^{1/2}$$
$$= 111.344^{1/2}$$
$$= 10.55 \text{ meters}$$

The distance is 10.55 meters along a straight line segment from the flashlight bulb to the point where the light beam's center strikes the ceiling.

QUIZ

Refer to the text in this chapter if necessary. A good score is eight correct. Answers are in the back of the book.

1. Fill in the blank to make the following statement true: "In the situation of Fig. 7-19, assuming that lines *L* and *M* both lie in plane *X*, line *N* constitutes a _____ to plane *X* at point *S*."
    A. bisector
    B. minor axis
    C. major axis
    D. normal

2. Imagine a triangle and its interior region in a Euclidean plane. Suppose that we "remove" the three line segments representing the triangle itself (i.e., the outer boundary of the region it encloses). How does this action affect the perimeter and area of the enclosed region?
    A. It does not affect either the perimeter or the area.
    B. It reduces the perimeter to zero, but does not change the area.
    C. It reduces both the perimeter and the area to zero.
    D. It renders both the perimeter and the area meaningless.

3. Whenever we talk or write about "the distance between a point and a line," "the distance between a line and a plane parallel to that line," or "the distance between two planes," we specify
    A. the longest possible distance unless we state otherwise.
    B. the shortest possible distance unless we state otherwise.

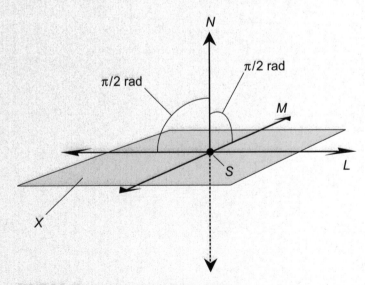

FIGURE 7-19 • Illustration for Quiz Question 1.

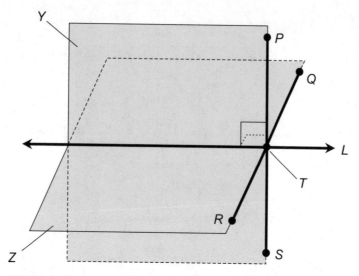

FIGURE 7-20 · Illustration for Quiz Question 4.

C. any distance between the shortest possible and the longest possible.

D. nothing whatsoever, unless we provide additional information.

4. Figure 7-20 shows two planes, *Y* and *Z*, which intersect along a line *L*. Line *PS* lies in plane *Y* and runs perpendicular to line *L*. Line *QR* lies in plane *Z* and runs perpendicular to line *L*. Point *T* lies at the intersection of lines *L*, *PS*, and *QR*. Based on this information, we can have absolute confidence that ∠*PTR* has the same measure as

A. ∠*STQ*.

B. ∠*RTS*.

C. ∠*QTP*.

D. All of the above

5. If two planes in Euclidean three-space share no points, then we can have complete confidence that the planes are

A. orthogonal.

B. skew.

C. parallel.

D. normal.

6. What's the smallest possible number of points that can uniquely define a plane in Euclidean three-space?

A. Two

B. Three

C. Four

D. Infinitely many

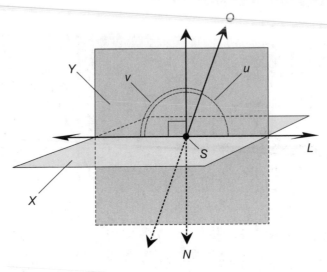

FIGURE 7-21 · Illustration for Quiz Question 8.

7. What's the largest possible number of intersecting lines that can uniquely define a plane in Euclidean three-space?

 A. Two
 B. Three
 C. Four
 D. Infinitely many

8. Figure 7-21 illustrates two planes *X* and *Y* that intersect along a line *L*. Line *N*, which lies in plane *Y*, runs normal to plane *X*. Suppose that line *O* lies in plane *Y* but does not run perpendicular to line *L*. All three lines *L*, *N*, and *O* intersect each other at point *S*. Consider the two angles of measures *u* and *v* with vertices at *S* shown. Based on this information, we know for sure that

 A. $u + v = \pi/4$.
 B. $u + v = \pi/2$.
 C. $u + v = \pi$.
 D. $u + v = 2\pi$.

9. In order for two lines in space to run parallel to each other, they must

 A. not intersect at any point.
 B. both lie in the same plane.
 C. not run askew relative to each other.
 D. All of the above

10. Consider a line in Euclidean three-space, and a point that doesn't lie on that line. What's the largest number of planes that can contain both the line and the point?

 A. One
 B. Two
 C. Three
 D. Infinitely many

chapter **8**

Surface Area and Volume

We can calculate the surface areas and volumes of various simple geometric solids in Euclidean three-space when we know the linear dimensions such as height, width, depth, or radius.

CHAPTER OBJECTIVES

In this chapter, you will

- Define and enumerate the most basic polyhedron types.
- Calculate polyhedron surface areas and volumes.
- Define and enumerate cones and cylinders.
- Calculate cone and cylinder surface areas and volumes.
- Define the sphere, ellipsoid, and torus.
- Calculate sphere, ellipsoid, and torus surface areas and volumes.

Straight-Edged Objects

Geometric solids with straight edges always have flat *faces*, also called *facets*, each of which forms a plane polygon. We call a 3D object of this sort a *polyhedron*.

The Tetrahedron

A polyhedron always has at least four faces. A four-faced polyhedron is called a *tetrahedron*. Each of the four faces constitutes a triangle. The tetrahedron has six edges where pairs of faces meet and four vertices where groups of edges meet. Any four specific points, as long as they don't all lie in a single plane, define a tetrahedron.

Surface Area of Tetrahedron

Figure 8-1 shows a tetrahedron whose height we call h; the shaded region portrays the base. We can calculate the surface area of the entire tetrahedron by adding up the interior areas of all four triangular faces. In the case of a *regular tetrahedron*, all six edges have the same length, so each face is an equilateral triangle. If the length of each edge of a regular tetrahedron equals s units, then we can calculate the surface area B of the whole object in square units (or units squared) with the formula

$$B = 3^{1/2}\, s^2$$

where $3^{1/2}$ represents the square root of 3, or approximately 1.732.

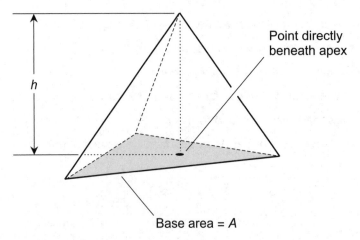

FIGURE 8-1 · A tetrahedron has four faces (including the base), six edges, and four vertices.

Volume of Tetrahedron

Imagine a tetrahedron whose base forms a triangle with area A, and whose height equals h as shown in Fig. 8-1. We can calculate the enclosed volume V of the solid in *cubic units* (or *units cubed*) using the formula

$$V = Ah/3$$

Pyramid

Figure 8-2 illustrates a *pyramid* whose height we call h. This figure has a square or rectangular base (shaded region) and four slanted faces above the base. In total the pyramid has five faces, eight edges where pairs of faces meet, and five vertices where groups of edges meet. If the base forms a perfect square and the *apex* (topmost vertex) lies directly above the point at the center of the base, then we have a *right square pyramid,* and each of the four of slanted faces constitutes an isosceles triangle. The well-known historical pyramids in Egypt are all of this type.

Surface Area of Pyramid

We can calculate the surface area of a pyramid by adding up the areas of all five of its faces (the four slanted faces plus the base). In the case of a right square pyramid where the length of each slanted edge, called the *slant height*, equals s units and the length of each edge of the base equals t units, the surface area B in square units is given by the formula

$$B = t^2 + 2t\ (s^2 - t^2/4)^{1/2}$$

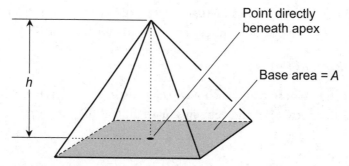

FIGURE 8-2 · A pyramid has five faces (including the base), eight edges, and five vertices.

Still Struggling

In the case of a general pyramid where the base doesn't form a square and/or the apex doesn't lie directly above the center of the base, the task of finding the surface area can sometimes present us with a tedious problem, because we must individually calculate the area of the base and each slanted face and then add the five polygons' areas up to get the total surface area.

Volume of Pyramid

Imagine a right square pyramid whose base is a square with area A, and whose height equals h as shown in Fig. 8-2. We can calculate the volume V of the pyramid in cubic units using the formula

$$V = Ah/3$$

We get cubic units when we multiply an area (in square units) by a linear dimension (in this case the height of the object, expressed in straight units).

TIP   *The pyramid volume formula holds true even if the base of the pyramid doesn't form a perfect square, and even if the apex point doesn't lie directly above the center of the base. In fact, the formula works for all types of pyramids, even grossly distorted ones, as long as we stay in Euclidean three-space.*

The Cube

Figure 8-3 illustrates a *cube*. This figure constitutes a *regular hexahedron* (six-sided polyhedron). It has 12 edges, all of which have identical length. Each of the six faces constitutes a square. The cube has eight vertices.

Surface Area of Cube

Imagine a cube whose edges each have length s, as shown in Fig. 8-3. We can find the surface area A of the cube in square units using the formula

$$A = 6s^2$$

We simply find the area of any single face and then multiply that area by 6 to obtain the total surface area of the solid.

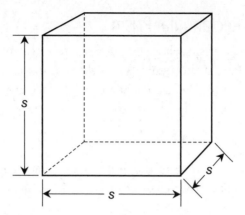

FIGURE 8-3 • A cube has six square faces and 12 edges of identical length.

Volume of Cube

Imagine a cube as defined above and in Fig. 8-3. We can calculate the volume V of the solid using the formula

$$V = s^3$$

Our job consists of nothing more than cubing (taking the third power of) the length of any one of the edges to get the volume in cubic units.

The Rectangular Prism

Figure 8-4 illustrates a geometric solid known as a *rectangular prism*. Each of the six faces is a rectangle. The figure has 12 edges. The edges of the entire object don't necessarily all have equal lengths, but the pair of edges at opposite sides of any given face does. The rectangular prism has eight vertices.

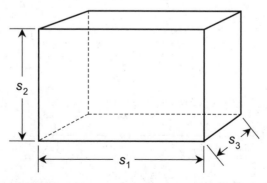

FIGURE 8-4 • A rectangular prism has six rectangular faces and 12 edges.

Surface Area of Rectangular Prism

Imagine a rectangular prism whose edges measure s_1, s_2, and s_3 units as shown in Fig. 8-4. (As we look at this figure, we might want to call s_1 the width, s_2 the height, and s_3 the depth.) We can calculate the surface area A of the prism in square units with the formula

$$A = 2s_1s_2 + 2s_1s_3 + 2s_2s_3$$

The process involves calculating the areas of each face and then adding the areas together. Note that any two opposite faces have identical areas.

Volume of Rectangular Prism

Imagine a rectangular prism as defined above and in Fig. 8-4. We can calculate the volume V of the enclosed solid in cubic units using the formula

$$V = s_1s_2s_3$$

We get cubic units when we multiply a linear dimension by another linear dimension and then multiply that result by a third linear dimension.

The Parallelepiped

We define a *parallelepiped* as a six-faced polyhedron in which each face constitutes a parallelogram, and opposite pairs of faces have identical size and shape as shown in Fig. 8-5. The figure has 12 edges and eight vertices. In this illustration, we call the smaller (acute or right) angles between the pairs of edges x, y, and z.

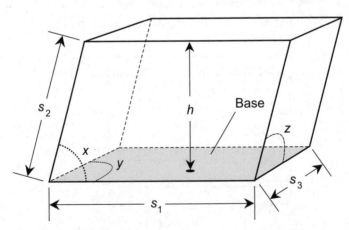

FIGURE 8-5 · A parallelepiped has six faces, all of which are parallelograms, and 12 edges.

As with the rectangular prism, the edges aren't necessarily all equally long, but the pair of edges at opposite sides of any given face has equal measure.

TIP *Every cube constitutes a rectangular prism whose edges all have equal length, and every rectangular prism constitutes a parallelepiped whose angles all measure 90° ($\pi/2$ rad). So, as things work out, some parallelepipeds are cubes (but most aren't).*

Surface Area of Parallelepiped

Imagine a parallelepiped with faces of lengths s_1, s_2, and s_3. Suppose that we call the angles between pairs of edges x, y, and z as shown in Fig. 8-5. We can determine the surface area A of the parallelepiped in square units using the formula

$$A = 2s_1s_2 \sin x + 2s_1s_3 \sin y + 2s_2s_3 \sin z$$

where $\sin x$ represents the sine of angle x, $\sin y$ represents the sine of angle y, and $\sin z$ represents the sine of angle z. As with the rectangular prism, any two opposite faces have identical areas. We determine the areas of all the faces first and then add those areas up to get the total surface area for the object.

Still Struggling

If you've forgotten how the sine function (and trigonometry in general) works, or if it otherwise baffles you, don't worry about the details. You can find the sine of any angle using a calculator, but be careful. If you express the angle in radians, you must set your calculator for radians. If you express the angle in degrees, you must set your calculator for degrees. (I've made that mistake more than once. It's easy for me to misadjust my computer's calculator when I work in radians, because the program uses degrees by default. On a few occasions, I forgot about that quirk until I got a result that obviously didn't make sense.)

Volume of Parallelepiped

Imagine a parallelepiped whose faces have lengths s_1, s_2, and s_3, and that has vertex angles of x, y, and z as shown in Fig. 8-5. Suppose that the height of the parallelepiped, as measured along a line normal to the base, equals h. We can find the volume V of the enclosed solid in cubic units by taking the product of the base area and the height according to the formula

$$V = hs_1s_3 \sin y$$

We must make sure that we take the sine of the correct angle when we use this formula! The variable y represents the acute angle between adjacent edges of the base.

 PROBLEM 8-1

Suppose that we want to paint the interior walls of a room. The room has the shape of a rectangular prism. The ceiling lies exactly 3.0 meters above the floor. The floor and the ceiling both measure exactly 4.2 meters by 5.5 meters. The room has two windows on its walls, the outer frames of which both measure 1.5 meters high by 1.0 meter wide. The outer frame of the doorway measures 2.5 meters high by 1.0 meter wide. We plan to cover all the walls with two coats of paint. A "paint guru" tells us that we can expect one liter of paint to cover exactly 20 square meters of wall area in a single coat. How much paint, in liters, will we need to completely do the job, without a single extra drop of paint to spare?

SOLUTION

Let's calculate our room's wall surface area, not including the door or the windows. Based on the information given, we can say that the rectangular prism formed by the edges between walls, floor, and ceiling measures 4.2 meters wide (dimension s_1 as portrayed in Fig. 8-4) by 3.0 meters high (dimension s_2 as portrayed in Fig. 8-4) by 5.5 meters deep (dimension s_3 as portrayed in Fig. 8-4). To find the total surface area A of the rectangular prism, in square meters including the windows and doorway, we use the formula

$$A = 2s_1s_2 + 2s_1s_3 + 2s_2s_3$$
$$= (2 \times 4.2 \times 3.0) + (2 \times 4.2 \times 5.5) + (2 \times 3.0 \times 5.5)$$
$$= 25.2 + 46.2 + 33.0$$
$$= 104.4 \text{ square meters}$$

Now we remember that the room has two windows, each one measuring 1.5 meters by 1.0 meter. Each window therefore takes away $1.5 \times 1.0 = 1.5$ square meters of area. The doorway measures 2.5 meters by 1.0 meter, so it takes away $2.5 \times 1.0 = 2.5$ square meters. The windows and doorway combined take away $1.5 + 1.5 + 2.5 = 5.5$ square meters of wall space. We must also subtract the combined areas of the floor and ceiling, neither of which we intend to paint. This quantity is the middle factor, 46.2, in the equation for the total surface area of the rectangular prism. We can now calculate the total wall area that we want to paint (let's call it A_w) as

$$A_w = (104.4 - 5.5) - 46.2$$
$$= 52.7 \text{ square meters}$$

We can expect a liter of paint to cover 20 square meters in a single coat. Therefore, we will need 52.7/20, or 2.635, liters of paint to coat the walls once. We'll need twice that much paint, or 5.27 liters, to finish the two-coat job without leaving any paint unused.

Circular Cones

In Euclidean three-space, a *circular cone* has a base that forms a perfect circle, and an apex point that lies outside the plane defined by that circle. The surface of any circular cone has the following components:

- The base circle
- All points inside the base circle that lie in its plane
- All line segments connecting the base circle (not including its interior) and the apex point

TIP   *The interior of the cone consists of the set of all points enclosed by the surface. In theory, we can include part, all, or none of the cone's surface when we talk about the solid. Usually, when we think of a solid cone, we imagine the entire surface (including the base) as well as the interior.*

Right Circular Cone

A *right circular cone* has a circular base, and an apex point that lies on a line normal to the plane of the base and that passes through the center of the base. Figure 8-6 shows an example. Line PQ runs normal to the plane containing the

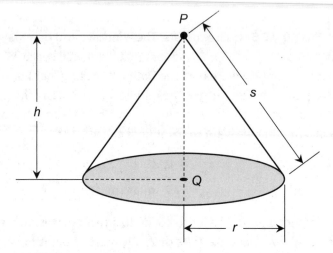

FIGURE 8-6 • A right circular cone.

base. Its length, representing the height of the cone, equals h. The radius of the circular base equals r. The cylinder's *slant height* is the distance s from the apex to any point on the edge of the base.

Surface Area of Right Circular Cone

Imagine a right circular cone as shown in Fig. 8-6. Let P represent the apex of the cone, and let Q represent the center of the base. Let r represent the radius of the base, let h represent the height, and let s represent the slant height. We can calculate the surface area S_1 of the cone, including the base, in square units with the formula

$$S_1 = \pi r^2 + \pi rs$$

Alternatively, we can use

$$S_1 = \pi r^2 + \pi r (r^2 + h^2)^{1/2}$$

The surface area S_2 of the cone, not including the base, is called the *lateral surface area* and is given by the formula

$$S_2 = \pi rs$$

We can also use

$$S_2 = \pi r (r^2 + h^2)^{1/2}$$

Volume of Right Circular Cone

Imagine a right circular cone as defined earlier and in Fig. 8-6. We can calculate the volume V of the entire solid in cubic units using the formula

$$V = \pi r^2 h/3$$

The quantity πr^2 represents the interior area of the circular base in square units. When we multiply square units by the height (a linear dimension), we get cubic units.

Surface Area of Frustum of Right Circular Cone

Imagine a right circular cone that's *truncated* ("chopped off") by a plane parallel to the base. We call the resulting object a *frustum* of the cone (Fig. 8-7). Let P represent the center of the circle defined by the truncation plane, and let Q represent the center of the base. Suppose that line segment PQ runs perpendicular to the base. Let r_1 represent the radius of the top circle (where we've "chopped" off the cone), let r_2 represent the radius of the base circle, let h represent the height of the object (the length of line segment PQ), and let s represent the slant height. If we don't know the slant height s, we can calculate the

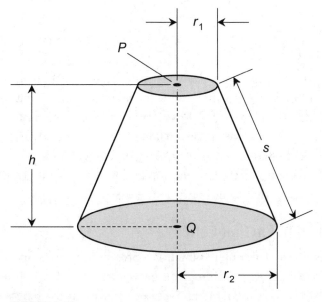

FIGURE 8-7 · Frustum of a right circular cone.

surface area S_1 of the object (including the base and the top) in square units with the rather messy formula

$$S_1 = \pi (r_1 + r_2)[s^2 + (r_2 - r_1)^2]^{1/2} + \pi (r_1^2 + r_2^2)$$

If we do know the slant height s, we can use the simpler formula

$$S_1 = \pi s (r_1 + r_2) + \pi (r_1^2 + r_2^2)$$

If we don't know the slant height s, we can calculate the surface area S_2 of the object (not including the base or the top) using the formula

$$S_2 = \pi(r_1 + r_2)[s^2 + (r_2 - r_1)^2]^{1/2}$$

Alternatively, if we do know s, we can use

$$S_2 = \pi s (r_1 + r_2)$$

Volume of Frustum of Right Circular Cone

Imagine a frustum of a right circular cone as defined earlier and as illustrated in Fig. 8-7. We can calculate the volume V of the enclosed solid in cubic units using the formula

$$V = \pi h (r_1^2 + r_1 r_2 + r_2^2)/3$$

The Slant Circular Cone

A *slant circular cone* has a base that constitutes a circle, and an apex point such that a normal line from the apex point to the plane containing the base does not pass through the center of the base. In "extreme slant circular cones," that line intersects the base plane on or outside the base circle. Figure 8-8 shows an example of a slant circular cone of the "extreme" type. Line segment PQ, which represents the height h, runs normal to plane X, which contains the base. The cone slants so much that Q lies outside the base.

Volume of Slant Circular Cone

Imagine a slant circular cone in which P represents the apex and Q represents a point in the plane X containing the base, such that line segment PQ runs perpendicular to X as shown in Fig. 8-8. Let h represent the height of the cone.

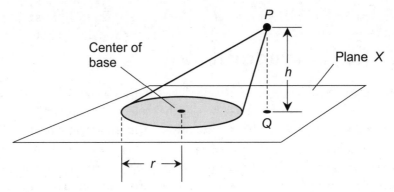

FIGURE 8-8 · A slant circular cone of the "extreme" type.

Let r represent the radius of the circular base. We can calculate the volume V of the solid in square units with the formula

$$V = \pi r^2 h/3$$

TIP   *The foregoing formula duplicates the one for the volume of a right circular cone. We can "push" the apex point P of a circular cone "sideways" as much as we want—even millions of times the radius of the base!—and as long as we don't alter the length of line segment PQ that runs normal to plane X, the volume of the enclosed solid will remain constant.*

Circular Cylinders

A *circular cylinder* has a base that forms a perfect circle, along with a circular top that has the same radius as the base and that lies in a plane parallel to the base. The cylinder itself comprises the following components:

- The base circle
- All points inside the base circle that lie in its plane
- The top circle
- All points inside the top circle that lie in its plane
- All line segments connecting the base circle and the top circle (not including their interiors) that run parallel to a line passing through the centers of both circles

Still Struggling

In the foregoing definition of a circular cylinder, the last "bulleted" item specifies line segments parallel to the line connecting the center of the base with the center of the top. When we impose this restriction, we can have absolute confidence that every such segment lies on the cylinder's surface, and none of the segments pass through the interior. When we combine all possible line segments of this sort (infinitely many of them exist), we get a curved surface that joins the base circle with the top circle. If we connect the base and top circles with any line segment that doesn't run parallel to the line connecting their centers, then that line segment runs through the interior of the cylinder, not along its outer surface.

TIP   *In theory, we can include part, all, or none of the surface of the cylinder when we define the entire solid. Normally, when we think of a solid cylinder, we think of the interior along with the entire surface including the base and the top.*

Right Circular Cylinder

A *right circular cylinder* has a circular base and a circular top. The base and the top lie in parallel planes. The center of the base and the center of the top lie at the ends of a line segment PQ that runs normal to both the plane containing the base and the plane containing the top, as shown in Fig. 8-9. The base circle

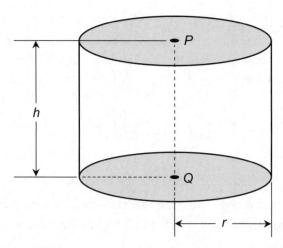

FIGURE 8-9 • A right circular cylinder.

and the top circle both have radius r. The length of line segment PQ equals the height h of the cylinder.

Surface Area of Right Circular Cylinder

Imagine a right circular cylinder where P represents the center of the top and Q represents the center of the base (Fig. 8-9). Let r represent the radii of the base and the top, and let h represent the height. We can calculate the surface area S_1 of the cylinder, including the base and the top, in square units with the formula

$$S_1 = 2\pi rh + 2\pi r^2$$
$$= 2\pi r(h + r)$$

The lateral surface area S_2 of the cylinder (not including the base or the top) is given by the simpler formula

$$S_2 = 2\pi rh$$

Volume of Right Circular Cylinder

Imagine a right circular cylinder as defined earlier and as shown in Fig. 8-9. We can calculate the volume V of the solid in cubic units with the formula

$$V = \pi r^2 h$$

The Slant Circular Cylinder

A *slant circular cylinder* has a circular base and a circular top. The base and the top lie in parallel planes. The center of the base and the center of the top lie along a line that does not run perpendicular to the planes that contain them (Fig. 8-10). The cylinder height h equals the distance between the plane containing the top and the plane containing the base, as determined along a line that runs normal to both planes. We represent the radii of the base and top circles as r.

Volume of Slant Circular Cylinder

Imagine a slant circular cylinder as defined earlier and as shown in Fig. 8-10. We can calculate the volume V of the solid in cubic units as

$$V = \pi r^2 h$$

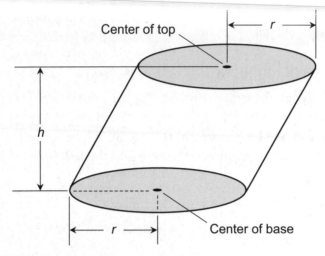

Center of top

r

h

r

Center of base

FIGURE 8-10 · A slant circular cylinder.

TIP   *The above formula duplicates the one for the volume of a right circular cylinder. We can "push" the top of a circular cylinder "sideways," and as long as we don't alter the cylinder's height, the volume of the enclosed solid will remain constant.*

PROBLEM 8-2

Imagine a cylindrical water tower that measures exactly 30 meters high and exactly 10 meters in radius. How many liters of water can it hold, assuming that we can fill up the entire interior with water? (One liter equals a *cubic decimeter:* the volume of a cube measuring 0.1 meter on an edge.) Round the answer off to the nearest liter.

SOLUTION

Let's use our formula to find the volume *V* in cubic meters in terms of the base or top radius *r* and the height *h*. That equation, once again, is

$$V = \pi r^2 h$$

We know that $r = 10$ and $h = 30$. If we consider $\pi = 3.141592654$ (that's more than enough decimal places for this calculation), then we can determine the interior volume of the cylinder as

$$V = 3.141592654 \times 10^2 \times 30$$
$$= 3.141592654 \times 100 \times 30$$
$$= 9424.777962 \text{ cubic meters}$$

One liter equals the volume of a cube that measures precisely 0.1 meter on an edge. That's $0.1 \times 0.1 \times 0.1$, or 0.001 (1/1000) of a cubic meter. Conversely, a cubic meter contains 1000 liters. We must therefore multiply the above-derived result by 1000 to get the answer in liters. When we do that, we get 9,424,777.962 liters. Rounding off to the nearest liter gives us a final answer of 9,424,778 liters.

PROBLEM 8-3

Imagine a circus tent that has the shape of a right circular cone. Suppose that its base diameter equals exactly 50 meters and the height at the center equals exactly 20 meters. How much canvas does the tent contain in terms of surface area? Express the answer to the nearest square meter. Assume that inside the tent, the floor is plain earth (not canvas).

SOLUTION

We can use the formula for the lateral surface area S of the right circular cone, not including the base, in terms of the radius r and the height h. Once again, that formula is

$$S = \pi r (r^2 + h^2)^{1/2}$$

We know that the tent's base diameter is precisely 50 meters. The radius equals half that span, so $r = 25$. We also know that $h = 20$. Let's consider $\pi = 3.141592654$. Then we have

$$S = 3.141592654 \times 25 \times (25^2 + 20^2)^{1/2}$$
$$= 3.141592654 \times 25 \times (625 + 400)^{1/2}$$
$$= 3.141592654 \times 25 \times 1025^{1/2}$$
$$= 3.141592654 \times 25 \times 32.01562119$$
$$= 2514.501009$$

The tent contains 2515 square meters of canvas, rounded off to the nearest square meter.

Other Solids

The realm of Euclidean three-space contains a tremendous variety of geometric solids that have curved surfaces throughout. Let's look at three of the most common such objects: the *sphere*, the *ellipsoid*, and the *torus*.

The Sphere

Consider a specific point P in Euclidean three-space. The surface of a sphere (call it S) comprises the collection of all points that lie at a specific distance or radius r from a defined point P. The interior of sphere S, including the surface, comprises the collection of all points whose distance from point P is less than or equal to r. The interior of sphere S, not including the surface, comprises the collection of all points whose distance from P is strictly less than r.

Surface Area of Sphere

Imagine a sphere S having radius r as shown in Fig. 8-11. We can calculate the surface area A of the sphere in square units with the formula

$$A = 4\pi r^2$$

Volume of Sphere

Imagine a sphere S as defined earlier and as illustrated in Fig. 8-11. We can calculate the volume V of the solid enclosed by the sphere in cubic units with the formula

$$V = 4\pi r^3/3$$

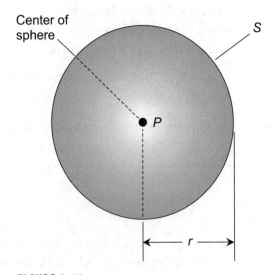

FIGURE 8-11 · A sphere.

TIP *The above formula for volume applies to the interior of sphere S, either includ-ing the surface or not including it, because the surface has zero volume. This same general concept holds true for volumes of all the other solids described in this chapter. We can take away part, or all, of the surface from a solid; any such action will have no effect on the enclosed volume. Think of the surface of any object, in mathematical terms, as an "infinitely thin shell." As such, it can possess no volume, no matter how many square units it has!*

The Ellipsoid

Let E represent a set of points that forms a *closed surface* (meaning that it has no "holes"; if we could fill it with air under pressure, none of the air would leak out). In this situation, E constitutes an ellipsoid if and only if, for any plane X that intersects E, the intersection between E and X forms a single point, a circle, or an ellipse.

Figure 8-12 shows an ellipsoid E with center point P and radii (also called *semi-axes*) r_1, r_2, and r_3, as we might specify them in a rectangular three-space coordi-nate system with P at the origin. If r_1, r_2, and r_3 are all equal, then E is a sphere. All spheres are ellipsoids, although plenty of nonspherical ellipsoids obviously exist.

Volume of Ellipsoid

Imagine an ellipsoid whose semiaxes measure r_1, r_2, and r_3 as shown in Fig. 8-12. We can calculate the volume V of the enclosed solid in cubic units with the formula

$$V = 4\pi r_1 r_2 r_3 / 3$$

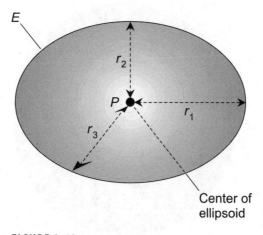

FIGURE 8-12 · An ellipsoid.

When we multiply three linear-unit quantities together (in this case the lengths of the three semiaxes of our ellipsoid), we get cubic units.

Still Struggling

Do you wonder why we don't mention a formula for calculating the surface area of an ellipsoid? There's a good reason: It's too complicated for this course! In order to precisely define the surface area of an ellipsoid in the general case where the three semiaxes can all differ from each other, we need to use *vector calculus*. Non-calculus formulas exist for approximating the surface area of a general ellipsoid, but they're messy and they rarely yield an exact answer. If you're curious about these approximation formulas, enter "surface area of an ellipsoid" into your favorite Internet search engine's phrase box and take things from there!

The Torus

Imagine a ray PQ and a small circle C centered on point Q with a radius less than half the distance between points P and Q. Suppose that we rotate the ray PQ, along with the small circle C centered at point Q, around its end point P, so that point Q describes a circle in a plane perpendicular to the small circle C. When we go through these maneuvers, the resulting collection of points in Euclidean three-space forms a torus. Figure 8-13 shows a torus T constructed

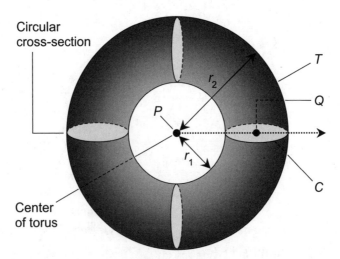

FIGURE 8-13 • A torus, also called a "donut."

in this fashion, with center point P. The inside radius equals r_1 and the outside radius equals r_2.

Surface Area of Torus

Imagine a torus with an inner radius of r_1 and an outer radius of r_2 as shown in Fig. 8-13. We can calculate the surface area A of the torus in square units with the formula

$$A = \pi^2 (r_2 + r_1)(r_2 - r_1)$$

Volume of Torus

Consider a torus T as defined earlier and as shown in Fig. 8-13. We can calculate the volume V of the enclosed solid in cubic units with the formula

$$V = \pi^2 (r_2 + r_1)(r_2 - r_1)^2/4$$

 **PROBLEM 8-4**

Suppose that we want to cover an American football field with an inflatable dome that takes the shape of a half-sphere. If the radius of the dome equals exactly 100 meters, what's the volume of air enclosed by the dome in cubic meters? Find the result to the nearest cubic meter.

SOLUTION

First, let's find the volume V of a sphere whose radius equals precisely 100 meters and then divide the result by 2. Consider $\pi = 3.141592654$. Using the formula with $r = 100$ gives us

$$V = 4\pi r^3/3$$
$$= (4 \times 3.141592654 \times 100^3)/3$$
$$= (4 \times 3.141592654 \times 1,000,000)/3$$
$$= 4,188,790.205$$

The volume enclosed by the dome equals half of this value. Calculating, we get

$$V/2 = 4,188,790.205/2$$
$$= 2,094,395.103$$

Rounding off to the nearest whole number, we get 2,094,395 cubic meters as the volume of air enclosed by the dome.

PROBLEM 8-5

Imagine that the dome over our American football field does not form a half-sphere, but instead constitutes a half-ellipsoid. Imagine that the height of the ellipsoid equals exactly 70 meters above its center point, which lies exactly in the center of the field. Suppose that the distance from the center of the field to either "far side" of the dome equals precisely 120 meters, and the distance from the center of the field to either "near side" of the dome equals precisely 90 meters. How many cubic meters of air does this dome enclose?

SOLUTION

First, let's consider the lengths of the semiaxes as $r_1 = 120$, $r_2 = 90$, and $r_3 = 70$. We can use the formula for the volume V of an ellipsoid with these radii, getting

$$V = 4\pi r_1 r_2 r_3/3$$
$$= (4 \times 3.141592654 \times 120 \times 90 \times 70)/3$$
$$= (4 \times 3.141592654 \times 756,000)/3$$
$$= 3,166,725.395$$

The volume enclosed by the dome equals half of this value. Calculating, we get

$$V/2 = 3,166,725.395/2$$
$$= 1,583,362.698$$

Rounding off to the nearest whole number, we get 1,583,363 cubic meters as the volume of air enclosed by the dome.

QUIZ

Refer to the text in this chapter if necessary. A good score is eight correct. Answers are in the back of the book.

1. **A cube constitutes a specific form of**
    A. four-faced polyhedron.
    B. six-faced polyhedron.
    C. eight-faced polyhedron.
    D. 12-faced polyhedron.

2. **If we double the lengths of two semiaxes in an ellipsoid while not changing the length of the third semiaxis, we increase the volume of the enclosed solid by a factor of**
    A. 2.
    B. 4.
    C. 8.
    D. 16.

3. **Figure 8-14 illustrates a right circular cone with dimensions precisely as indicated. What's the lateral surface area of this object (the surface area of the conical portion only, not including the base), rounded off to the nearest hundredth of a square unit?**
    A. 11.33 square units
    B. 17.61 square units
    C. 22.65 square units
    D. 35.22 square units

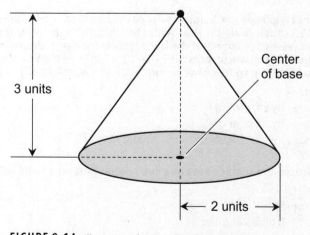

FIGURE 8-14 · Illustration for Quiz Questions 3 and 4.

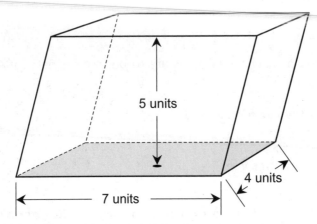

5 units

4 units

7 units

FIGURE 8-15 · Illustration for Quiz Questions 5 and 6.

4. **What's the exact volume of the solid enclosed by the right circular cone shown in Fig. 8-14, assuming that the object has precisely the dimensions indicated?**
   A. 2π cubic units
   B. 3π cubic units
   C. 4π cubic units
   D. 6π cubic units

5. **What's the volume of the solid enclosed by the parallelepiped shown in Fig. 8-15, given a width of exactly 7 units, a slant depth of exactly 5 units, and a height of exactly 5 units as indicated? Round off the answer to the nearest cubic unit.**
   A. 140 cubic units
   B. 35 cubic units
   C. 28 cubic units
   D. We need more information to answer this question.

6. **Suppose that we triple the height of the parallelepiped shown in Fig. 8-15, from 5 units to 15 units. If we do that while leaving the base dimensions at 7 by 4 units, and we also ensure that the base retains the same shape, what happens to the volume of the enclosed solid? (Here's a hint: We don't have to know the actual volume of the solid, either before or after the height-tripling action.)**
   A. It triples.
   B. It increases by a factor of 9.
   C. It increases by a factor of 27.
   D. It increases by a factor equal to the square root of 27.

7. **If we quadruple the surface area of a cube, what happens to its volume?**
   A. It doubles.
   B. It quadruples.
   C. It becomes 8 times as great.
   D. It becomes 16 times as great.

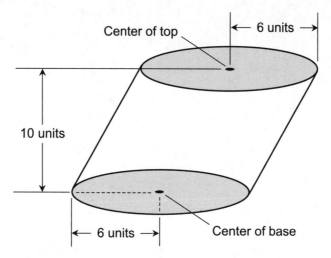

Center of top

6 units

10 units

6 units   Center of base

FIGURE 8-16 · Illustration for Quiz Questions 9 and 10.

8. **If we quadruple the surface area of a sphere, what happens to its volume?**

   A. It doubles.

   B. It quadruples.

   C. It becomes 8 times as great.

   D. It becomes 16 times as great.

9. **Figure 8-16 illustrates a slant circular cylinder with a radius of exactly 6 units and a height of exactly 10 units. What's the volume of the enclosed solid, rounded off to the nearest cubic unit?**

   A. 188 cubic units

   B. 377 cubic units

   C. 1131 cubic units

   D. We need more information to answer this question.

10. **Suppose that we triple the radius of the slant circular cylinder shown in Fig. 8-16, from 6 to 18 units. What happens to the volume of the enclosed solid? (Here's a hint: We don't have to know the actual volume of the solid, either before or after the radius-tripling action.)**

   A. It triples.

   B. It increases by a factor of 9.

   C. It increases by a factor of 27.

   D. It increases by a factor of 81.

chapter 9

Vectors and Cartesian Three-Space

We can define *Cartesian three-space*, also called *rectangular three-space* or *xyz-space*, on the basis of three real-number lines that intersect at a common origin point. At the origin, each number line runs perpendicular to the other two, so we can graphically relate one variable to both of the others. Most three-dimensional (3D) graphs show up in this system as lines, curves, or surfaces.

CHAPTER OBJECTIVES

In this chapter, you will

- Review the fundamentals of the sine, cosine, and tangent functions.
- Define vectors in Cartesian two-space.
- Learn how to add and "multiply" vectors in two-space.
- Construct a Cartesian three-space coordinate system.
- Define vectors in Cartesian three-space.
- Learn how to add and "multiply" vectors in three-space.
- Define and work with planes and lines in Cartesian three-space.

A Taste of Trigonometry

Before we proceed further, let's review a few principles of basic trigonometry. In particular, let's look at angle notation and the *sine*, *cosine*, and *tangent* functions.

It's Greek to Us

Mathematicians and scientists often use Greek letters to represent angles. The most common symbol for an angle is an italic, lowercase Greek letter *theta* (pronounced "THAY-tuh"). It looks like an italic numeral zero with a horizontal line inside (θ).

When writing about two different angles, we need to use another Greek letter along with θ. Most mathematicians prefer the italic, lowercase letter *phi* (pronounced "FIE" or "FEE"). It looks like an italic lowercase English letter o with a forward slash passing through (ϕ).

TIP *Numeric or variable subscripts are sometimes used with Greek symbols for angles, so you can expect to occasionally encounter angles denoted as θ_1, θ_2, θ_3, and so on, or as θ_x, θ_y, θ_z, and so on.*

The Unit Circle

Consider a circle in the Cartesian *xy*-plane with the following equation. It's the simplest possible circle, expressible as

$$x^2 + y^2 = 1$$

This equation represents a *unit circle*. That means it's centered at the origin point (0,0) on the coordinate plane and has a radius of 1 unit. Let θ represent an angle whose vertex point lies at the origin, and that we express going around counterclockwise from the *x* axis as shown in Fig. 9-1. Imagine that θ defines the direction of a ray that starts out at the origin and passes through the unit circle, intersecting the circle at the point $P = (x_0, y_0)$. We can define three trigonometric functions, called *circular functions*, of the angle θ.

The Sine Function

In Fig. 9-1, let *OP* represent the ray that emerges from the origin (point O) and passes through point *P* on the unit circle. Suppose that this ray starts out pointing

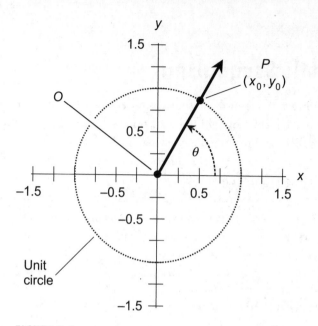

FIGURE 9-1 · The unit circle constitutes the basis for the trigo-nometric functions.

exactly along the x axis. Then it starts to rotate counterclockwise, always keeping its back-end point at the origin, as if the origin were a hinge or pivot. As the ray rotates, the point P, represented by coordinates (x_0, y_0), revolves around the unit circle.

Imagine what happens to the value of y_0 during one complete rotation of ray OP, starting out along the x axis and eventually returning there:

- The ray starts out such that $y_0 = 0$; then y_0 increases until it attains a value of 1 after P has gone 90° or $\pi/2$ rad around the circle ($\theta = 90° = \pi/2$ rad).
- After that, y_0 begins to decrease, getting back to a value of 0 when P has gone 180° or π rad around the circle ($\theta = 180° = \pi$ rad).
- As P continues counterclockwise, y_0 keeps decreasing until, at $\theta = 270° = 3\pi/2$ rad, the value of y_0 reaches its minimum of -1.
- After that, y_0 increases again until, when P has gone completely around the circle, it returns to the value of 0 for $\theta = 360° = 2\pi$ rad.

We define the value of y_0, for any particular angle θ, as the sine of θ. The sine function is abbreviated sin, so we can write

$$\sin \theta = y_0$$

Still Struggling

If you've always thought of the sine, cosine, and tangent functions as relating the relative side lengths of right triangles, now is the time to revolutionize your thinking! Yes, these three functions can and do describe the dimensions of right triangles, just as you've learned in other courses. However, the *unit-circle model* improves on the *right-triangle model* in at least two ways. First, the unit-circle model allows you to define negative as well as positive values for the trigonometric functions, and you can't do that with triangles. Second, the unit-circle model allows for angles measuring less than 0° (0 rad) or more than 90° ($\pi/2$ rad), while the right-triangle model forces you to stay within that range.

The Cosine Function

Look again at Fig. 9-1. Imagine, once again, a ray OP from the origin outward through point P on the circle, pointing along the x axis and then rotating in a counterclockwise direction:

- The ray starts out such that $x_0 = 1$; then x_0 decreases until it attains a value of 0 after P has gone 90° or $\pi/2$ rad around the circle ($\theta = 90° = \pi/2$ rad).

- After that, x_0 continues to decrease, reaching a minimum value of −1 when P has gone 180° or π rad around the circle ($\theta = 180° = \pi$ rad).

- As P continues counterclockwise, x_0 increases until, at $\theta = 270° = 3\pi/2$ rad, it gets back up to 0.

- After that, x_0 continues to increase until, when P has gone completely around the circle, it returns to the value of 1 for $\theta = 360° = 2\pi$ rad.

We define the value of x_0, for any particular angle θ, as the cosine of θ. The cosine function is abbreviated cos, so we can write

$$\cos \theta = x_0$$

The Tangent Function

Once again, refer to Fig. 9-1. We can define the tangent (abbreviated tan) of an angle θ using the same ray OP and the same point $P = (x_0, y_0)$ as we have done with the sine and cosine functions. The definition is

$$\tan \theta = y_0/x_0$$

Because we already know that $\sin \theta = y_0$ and $\cos \theta = x_0$, we can express the tangent function in terms of the sine and the cosine with the formula

$$\tan \theta = \sin \theta / \cos \theta$$

TIP *The tangent function is interesting because, unlike the sine and cosine functions, it becomes* **singular** *("blows up" in a positive or negative direction) at certain values of θ. Whenever $x_0 = 0$, the denominator of either quotient above becomes zero. Mathematicians don't allow, or even attempt to define, division by zero, so we cannot define the value of the tangent function for any angle θ such that $\cos \theta = 0$. Such angles include all possible odd-integer multiples of 90° ($\pi/2$ rad).*

PROBLEM 9-1

What's the tangent of 45°? Don't do any calculations. You can infer the result without having to write down a single numeral and without using a calculator.

SOLUTION

Draw a diagram of a unit circle, such as the one in Fig. 9-1, and place ray *OP* such that it subtends an angle of 45° with respect to the *x* axis. That's the angle for which you want to find the tangent. Note that ray *OP* also subtends an angle of 45° with respect to the *y* axis, because the *x* and *y* axes run perpendicular (they're oriented at 90° with respect to each other), and 45° equals half of 90°. Every point on the ray *OP*, including (x_0, y_0), lies equally distant from the *x* and *y* axes. It follows that x_0 and y_0 must have the same value, and neither of them is zero. You must conclude that the ratio of y_0 to x_0 equals 1, because any nonzero number divided by itself equals 1. According to the definition of the tangent function, therefore, $\tan 45° = 1$.

Vectors in the Cartesian Plane

A *vector* expresses a quantity with two independent properties: *magnitude* and *direction*. We define the direction, also called *orientation*, in the sense of a ray; it "points" somewhere. We can use vectors to represent physical variables such as displacement, velocity, and acceleration. Mathematicians and scientists usually

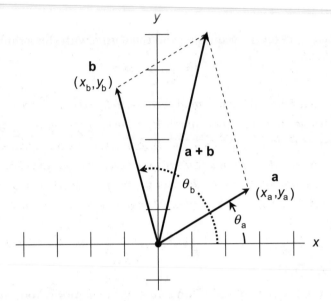

FIGURE 9-2 • Two vectors in the Cartesian plane. We can add them geometrically using the parallelogram method.

denote vectors using boldface letters of the alphabet. For example, in the Cartesian xy-plane, we can portray vectors **a** and **b** as rays from the origin $(0,0)$ to points (x_a,y_a) and (x_b,y_b), as shown in Fig. 9-2.

Equivalent Vectors

Occasionally, we'll encounter a vector that begins at a point other than the origin $(0,0)$. In order for the following formulas to hold, we must convert (or *reduce*) such a vector to the so-called *standard form*, such that it begins at the origin. We can carry out this task by subtracting the coordinates (x_0,y_0) of the starting point from the coordinates (x_1,y_1) of the end point. For example, if a vector **a**∗ starts at $(3,-2)$ and ends at $(1,-3)$, it reduces to an *equivalent vector* **a** in standard form as follows:

$$\mathbf{a} = \{(1-3),[-3-(-2)]\}$$
$$= [(1-3),(-3+2)]$$
$$= (-2,-1)$$

We define any vector **a**∗ that runs parallel to **a**, and that has the same length and the same direction (or orientation) as **a**, as *equal* to vector **a**.

TIP *We can define a vector solely on the basis of its magnitude and its direction (or orientation). Neither of these two properties depends on the location of the originating point.*

Magnitude

We can calculate the magnitude (also called the *length, intensity,* or *absolute value*) of vector **a**, written |a| or *a*, in the Cartesian plane by using a formula resembling the theorem of Pythagoras for right triangles:

$$|a| = (x_a^2 + y_a^2)^{1/2}$$

The vector magnitude equals the distance from the originating (or back-end) point to the terminating (or far-end) point.

Direction

The direction of vector **a**, written dir **a**, equals the angle θ_a that vector **a** subtends as expressed going around counterclockwise from the positive *x* axis in the Cartesian plane:

$$\text{dir } a = \theta_a$$

The tangent of the angle θ_a equals y_a/x_a. Therefore, θ_a equals the *inverse tangent,* also called the *arctangent* (abbreviated arctan or tan^{-1}) of y_a/x_a. We have

$$\text{dir } a = \theta_a$$
$$= \arctan (y_a/x_a)$$
$$= \tan^{-1} (y_a/x_a)$$

By convention, we should always reduce any angle θ_a to a value that's at least zero, but less than one full counterclockwise revolution. That is, we should always have

$$0° \leq \theta_a < 360°$$

if we express θ_a in degrees, or

$$0 \leq \theta_a < 2\pi$$

if we express θ_a in radians.

TIP *If we ever encounter an angle that doesn't fall within the above-defined range, we can reduce it to its conventional value (within that range) by adding or subtracting some integer multiple of 360° (2π rad).*

Sum

We can determine the sum of two vectors **a** and **b**, where $\mathbf{a} = (x_a, y_a)$ and $\mathbf{b} = (x_b, y_b)$, by adding their components individually using the formula

$$\mathbf{a} + \mathbf{b} = [(x_a + x_b), (y_a + y_b)]$$

We can also sum two vectors **a** and **b** geometrically by constructing a parallelogram with **a** and **b** forming a pair of adjacent sides. When we do that, **a** + **b** lies along the diagonal of the parallelogram as shown in Fig. 9-2 on page 224. Some people call this scheme the *parallelogram method* of vector addition.

Multiplication by Scalar

When we want to multiply a vector by a *scalar* (an ordinary real number), we multiply the x and y components of the vector individually by that scalar. If we have a vector $\mathbf{a} = (x_a, y_a)$ and a scalar k, then

$$k\mathbf{a} = \mathbf{a}k$$
$$= k\,(x_a, y_a)$$
$$= (kx_a, ky_a)$$

Still Struggling

Multiplication by a scalar changes the length of a vector, but not the orientation of the line along which it runs. If the scalar is positive, the direction of the product vector is the same as that of the original vector. If the scalar is negative, the direction of the product vector is opposite that of the original vector. If the scalar is zero, the product vector vanishes altogether.

Dot Product

Consider two vectors $\mathbf{a} = (x_a, y_a)$ and $\mathbf{b} = (x_b, y_b)$. We define the *dot product*, also known as the *scalar product* and written $\mathbf{a} \bullet \mathbf{b}$, of two vectors **a** and **b** as the real number (or scalar quantity) that we get when we use the formula

$$\mathbf{a} \bullet \mathbf{b} = x_a x_b + y_a y_b$$

PROBLEM 9-2

What's the sum of the two vectors a = (3,–5) and b = (2,6) in the Cartesian plane?

SOLUTION

We add the x and y components independently, obtaining

$$a + b = [(3 + 2),(-5 + 6)]$$
$$= (5,1)$$

PROBLEM 9-3

What's the dot product of the two Cartesian-plane vectors a = (3,–5) and b = (2,6)?

SOLUTION

Using the formula given above for the dot product, we get

$$a \bullet b = (3 \times 2) + (-5 \times 6)$$
$$= 6 + (-30)$$
$$= -24$$

PROBLEM 9-4

What happens if we reverse the order of a dot product? Does the value change? If so, how?

SOLUTION

No, the value does not change. The dot product of two vectors does not depend on the order in which we "dot-multiply" them. We can prove this fact in the general case using the formula from above. Let a = (x_a, y_a) and b = (x_b, y_b). First consider the dot product of a and b (pronounced "a dot b"):

$$a \bullet b = x_a x_b + y_a y_b$$

Now consider the dot product b $\bullet$ a:

$$b \bullet a = x_b x_a + y_b y_a$$

Because ordinary multiplication is commutative (the order in which we multiply the factors doesn't matter), we can convert the above formula to

$$b \bullet a = x_a x_b + y_a y_b$$

Now we can see that the quantity $x_a x_b + y_a y_b$ represents the expansion of a • b. Therefore, for any two vectors a and b, we always have

$$a \bullet b = b \bullet a$$

Three Number Lines

Figure 9-3 illustrates the simplest possible set of *rectangular 3D coordinates*. All three number lines have equal increments. (This drawing is a perspective illustration, so the increments on the z axis appear distorted. A true 3D rendition

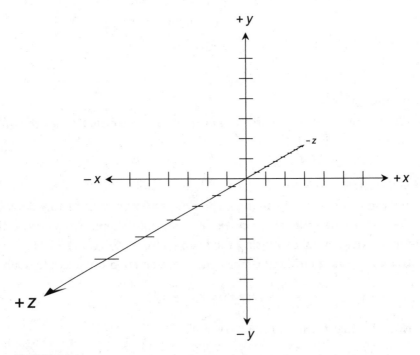

Each division equals 1 unit

FIGURE 9-3 • Cartesian three-space, also called *xyz*-space.

would have the positive *z* axis perpendicular to the page.) The three number lines intersect at their zero points. In this particular rendition:

- We call the horizontal (right-and-left) axis the *x* axis.
- We call the vertical (up-and-down) axis the *y* axis.
- We call the page-perpendicular (in-and-out) axis the *z* axis.

Still Struggling

In our portrayal of rectangular 3D coordinates, the positive *x* axis runs from the origin toward the viewer's right, and the negative *x* axis runs toward the left. The positive *y* axis runs upward, and the negative *y* axis runs downward. The positive *z* axis comes "out of the page," and the negative *z* axis extends "back behind the page." However, you'll find variations in some texts. You might see the positive *x* axis running from the origin toward the right, the negative *x* axis running toward the left, the positive *y* axis running "behind the page away from you," the negative *y* axis running "out of the page toward you," the positive *z* axis running vertically upward, and the negative *z* axis running vertically downward. However you see the axes portrayed, their *relative orientation* remains the same in all texts—unless an author or illustrator has made a mistake!

Ordered Triples as Points

Figure 9-4 shows two specific points, called *P* and *Q*, plotted in Cartesian three-space. The coordinates of point *P* are (−5,−4,3), and the coordinates of point Q are (3,5,−2). We denote point locations as *ordered triples* in the form (x,y,z), where the first number represents the value on the *x* axis, the second number represents the value on the *y* axis, and the third number represents the value on the *z* axis. The word "ordered" tells us that the order, or sequence, in which the numbers are listed is important. The ordered triple (1,2,3) is not the same as any of the ordered triples (1,3,2), (2,1,3), (2,3,1), (3,1,2), or (3,2,1), even though all of the triples contain the same three numbers.

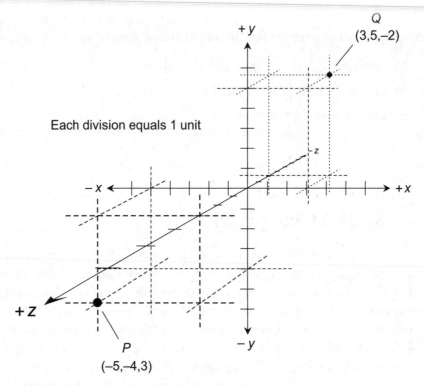

FIGURE 9-4 • Two points in Cartesian three-space.

TIP   *When you write an ordered triple, don't put any spaces after the commas, as you would do in the notation of a set, sequence, or list of numbers. Run the whole expression together without any spaces, and always remember to enclose it in parentheses.*

Variables and Origin

In Cartesian three-space, we usually have two independent-variable coordinate axes and one dependent-variable axis. The x and y axes represent independent variables, while the z axis represents a dependent variable whose value is affected by both the x and the y values.

In some scenarios, two of the variables are dependent and only one is independent. Most often, the independent variable in such cases is x.

Rarely, you'll come across a situation in which none of the values depends on either of the other two, or when a correlation *without any mathematical relation* exists among the values of two or all three of the variables. Plots of this sort usually appear as "swarms of points," representing the results of observations, or values "predicted" by a scientific theory.

Distance between Points

Consider two different points $P = (x_0, y_0, z_0)$ and $Q = (x_1, y_1, z_1)$ in Cartesian three-space. We can calculate the distance d between these two points using the formula

$$d = [(x_1 - x_0)^2 + (y_1 - y_0)^2 + (z_1 - z_0)^2]^{1/2}$$

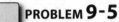 **PROBLEM 9-5**

What's the distance between the points $P = (-5, -4, 3)$ and $Q = (3, 5, -2)$ illustrated in Fig. 9-4? Express the answer rounded off to three decimal places.

 **SOLUTION**

Let's plug the coordinate values into the distance equation, where

$$x_0 = -5$$
$$x_1 = 3$$
$$y_0 = -4$$
$$y_1 = 5$$
$$z_0 = 3$$
$$z_1 = -2$$

When we grind out the arithmetic and round the final result off to three decimal places, we get

$$d = [(x_1 - x_0)^2 + (y_1 - y_0)^2 + (z_1 - z_0)^2]^{1/2}$$
$$= \{[3 - (-5)]^2 + [5 - (-4)]^2 + (-2 - 3)^2\}^{1/2}$$
$$= [8^2 + 9^2 + (-5)^2]^{1/2}$$
$$= (64 + 81 + 25)^{1/2}$$
$$= 170^{1/2}$$
$$= 13.038$$

Vectors in Cartesian Three-Space

A *vector* in Cartesian three-space resembles a vector in the Cartesian plane, except that a three-space vector has more "freedom" in terms of possible directions or orientations. This expanded scenario makes the expression of vector direction in 3D more complicated than it is in 2D. If you like vector analysis, you'll also find 3D vector arithmetic more interesting than two-dimensional (2D) vector arithmetic.

Equivalent Vectors

In Cartesian three-space, we can denote two vectors (let's call them **a** and **b**) as arrow-tipped line segments from the origin (0,0,0) to points (x_a, y_a, z_a) and (x_b, y_b, z_b), as shown in Fig. 9-5. This rendition, like all of the three-space drawings in this chapter, is a perspective illustration. Both vectors in this example point in directions on our side of the plane containing the page. In a true 3D model, both of them would "stick up out of the paper."

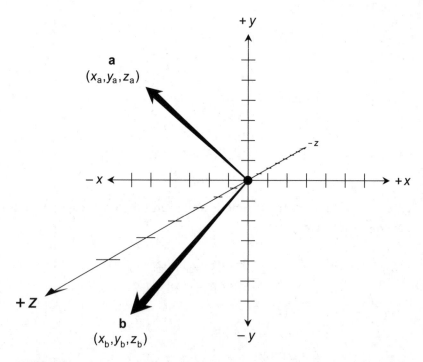

FIGURE 9-5 • Vectors in *xyz*-space. This is a perspective drawing; both vectors point in directions on our side of the plane containing the page.

In Fig. 9-5, both vectors **a** and **b** have their back-end points at the origin $(0,0,0)$. This situation represents the standard form of a vector in any coordinate system. In order for the following formulas to hold, we must express all vectors in standard form. If a given vector is not in standard form, we can convert it to that form by subtracting the coordinates (x_0,y_0,z_0) of the starting point from the coordinates (x_1,y_1,z_1) of the terminating point. For example, if a vector **a*** starts at $(4,7,0)$ and ends at $(1,-3,5)$, it reduces to an equivalent vector **a** in standard form as follows:

$$\mathbf{a} = [(1-4),(-3-7),(5-0)]$$
$$= (-3,-10,5)$$

TIP   *By definition, if some vector a* runs parallel to a, has the same length as a, and points in the same direction as a, then a* = a. Similarly, if some vector b* runs parallel to b, has the same length as b, and points in the same direction as b, then b* = b. As in the 2D case, we define a 3D vector solely on the basis of its magnitude and its direction. Neither of these two properties depends on the location of the originating or back-end point.*

Defining the Magnitude

When the back-end point of a vector **a** lies at the coordinate origin, we can find the magnitude of **a**, written |a| or a, using a 3D extension of the Pythagorean theorem for right triangles, as follows:

$$|a| = (x_a^2 + y_a^2 + z_a^2)^{1/2}$$

The magnitude of any vector **a** in standard form equals the distance of the terminating point from the coordinate origin. Note that the above formula is the distance formula for the specific case of two points $(0,0,0)$ and (x_a,y_a,z_a).

Direction Angles and Cosines

We can define the direction of a vector **a** in standard form by specifying the angles θ_x, θ_y, and θ_z that the vector **a** subtends relative to the positive x, y, and z axes, respectively, as shown in Fig. 9-6. We call these three angles, expressed in combination as the ordered triple $(\theta_x,\theta_y,\theta_z)$, the *direction angles* for the vector **a**.

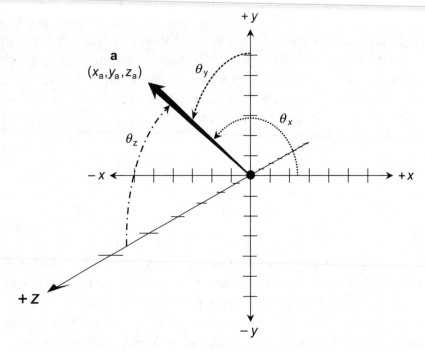

FIGURE 9-6 • Direction angles of a vector in *xyz*-space. This is another perspective drawing; the vector points in a direction on our side of the plane containing the page.

Sometimes, mathematicians will talk about the cosines of the direction angles to define the direction of a particular vector **a** in 3D space. We call such values the *direction cosines* of **a** and denote them with lower case Greek letters alpha (α), beta (β), and gamma (γ), as follows:

$$\text{dir } \mathbf{a} = (\alpha, \beta, \gamma)$$

where

$$\alpha = \cos \theta_x$$
$$\beta = \cos \theta_y$$
$$\gamma = \cos \theta_z$$

For any vector **a** in Cartesian three-space, the sum of the squares of the direction cosines always equals 1. That is,

$$\alpha^2 + \beta^2 + \gamma^2 = 1$$

We can also express this fact using the equation

$$\cos^2 \theta_x + \cos^2 \theta_y + \cos^2 \theta_z = 1$$

where the expression $\cos^2 \theta$ means $(\cos \theta)^2$.

Sum

We can calculate the sum of two vectors $\mathbf{a} = (x_a, y_a, z_a)$ and $\mathbf{b} = (x_b, y_b, z_b)$ in standard form by adding their components individually with the formula

$$\mathbf{a} + \mathbf{b} = [(x_a + x_b), (y_a + y_b), (z_a + z_b)]$$

This sum can, as in the 2D case, be found geometrically by constructing a parallelogram with $\mathbf{a}$ and $\mathbf{b}$ as adjacent sides. The sum $\mathbf{a} + \mathbf{b}$ corresponds to the diagonal of the parallelogram. Figure 9-7 shows an example. (The parallelogram appears distorted because of perspective effects.)

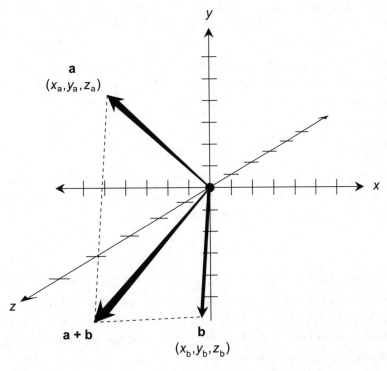

FIGURE 9-7 · We can add vectors in *xyz*-space using the "parallelogram method." This is a perspective drawing, so the parallelogram appears distorted.

Multiplication by Scalar

In 3D Cartesian coordinates, let's define vector **a** using the coordinates (x_a, y_a, z_a) when reduced to standard form. Suppose that we multiply this vector **a** by a positive real scalar k. We can express the product using the equation

$$k\mathbf{a} = \mathbf{a}k$$
$$= k\,(x_a, y_a, z_a)$$
$$= (kx_a, ky_a, kz_a)$$

If we multiply the vector **a** by a negative real scalar $-k$, then we have

$$-k\mathbf{a} = \mathbf{a}(-k)$$
$$= -k\,(x_a, y_a, z_a)$$
$$= (-kx_a, -ky_a, -kz_a)$$

Let's represent the direction angles for **a** as the ordered triple $(\theta_{xa}, \theta_{ya}, \theta_{za})$. The direction angles for vector $k\mathbf{a}$ coincide with those for **a**, that is, $(\theta_{xa}, \theta_{ya}, \theta_{za})$. However, the direction angles for $-k\mathbf{a}$ all differ by 180° (π rad) from those for **a** and $k\mathbf{a}$, indicating that $-k\mathbf{a}$ points in precisely the opposite direction from **a** and $k\mathbf{a}$. We can obtain the direction angles for $-k\mathbf{a}$ by adding or subtracting 180° (π rad) to or from each of the direction angles for $k\mathbf{a}$, so that the resulting angles are all positive but less than 360° (2π rad).

Dot Product

The *dot product*, also known as the *scalar product* and written **a** • **b**, of two Cartesian three-space vectors $\mathbf{a} = (x_a, y_a, z_a)$ and $\mathbf{b} = (x_b, y_b, z_b)$ in standard form equals a real number given by the formula

$$\mathbf{a} \bullet \mathbf{b} = x_a x_b + y_a y_b + z_a z_b$$

You can also calculate the dot product from the vector magnitudes |**a**| and |**b**| along with the angle θ between **a** and **b** as measured going counterclockwise in the plane containing them both. Multiply the two vector magnitudes by each other, and then multiply the result by the cosine of the angle between the vectors. You can use the formula

$$\mathbf{a} \bullet \mathbf{b} = |\mathbf{a}|\,|\mathbf{b}|\,\cos\theta$$

Cross Product

The *cross product*, also known as the *vector product* and written $\mathbf{a} \times \mathbf{b}$, of two vectors $\mathbf{a} = (x_a, y_a, z_a)$ and $\mathbf{b} = (x_b, y_b, z_b)$ in standard form is a third vector that runs perpendicular to the plane containing both $\mathbf{a}$ and $\mathbf{b}$. Let θ represent the angle between vectors $\mathbf{a}$ and $\mathbf{b}$ as measured going counterclockwise in the plane containing them both, as shown in Fig. 9-8. You can calculate the magnitude of the cross-product vector $\mathbf{a} \times \mathbf{b}$ using the formula

$$|\mathbf{a} \times \mathbf{b}| = |\mathbf{a}|\,|\mathbf{b}|\sin\theta$$

In the example shown, $\mathbf{a} \times \mathbf{b}$ points upward at a right angle to the plane containing the two vectors $\mathbf{a}$ and $\mathbf{b}$.

If $0° < \theta < 180°$ (0 rad $< \theta <$ π rad), you can use a trick called the *right-hand rule* to ascertain the direction of $\mathbf{a} \times \mathbf{b}$. Curl the fingers of your right hand in the rotational sense that you want to express θ, the angle starting in the direction of $\mathbf{a}$ and ending up in the direction of $\mathbf{b}$. (Make sure that you don't accidentally go in the rotational sense from $\mathbf{b}$ to $\mathbf{a}$!) Once you've got your hand in the correct position, extend your thumb straight out as if you're "hitchhiking." Your thumb will then point in the direction of $\mathbf{a} \times \mathbf{b}$.

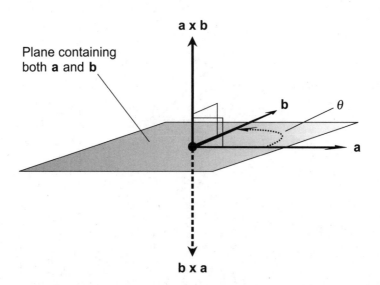

FIGURE 9-8 · The vector $\mathbf{b} \times \mathbf{a}$ has the same magnitude as vector $\mathbf{a} \times \mathbf{b}$, but points in the opposite direction. Both cross-product vectors point in directions perpendicular to the plane containing the two original vectors.

TIP   *When $180° < \theta < 360°$ (π rad $< \theta < 2\pi$ rad), the cross-product vector reverses direction because its magnitude (as determined by the above formula) turns out negative. We can see this fact when we note that $\sin \theta > 0$ when $0° < \theta < 180°$ (0 rad $< \theta < \pi$ rad), but $\sin \theta < 0$ when $180° < \theta < 360°$ (π rad $< \theta < 2\pi$ rad). When a formula gives us a negative vector magnitude going in a certain direction, we should think of it as an equivalent positive magnitude (i.e., −1 times the negative magnitude) in the opposite direction!*

Unit Vectors

Any vector **a**, reduced to standard form so its starting point lies at the origin, ends up at some point (x_a, y_a, z_a). We can break any such vector **a** down into the sum of three mutually perpendicular vectors, each of which lies along one of the coordinate axes as shown in Fig. 9-9:

$$\mathbf{a} = (x_a, y_a, z_a)$$
$$= (x_a, 0, 0) + (0, y_a, 0) + (0, 0, z_a)$$
$$= x_a(1, 0, 0) + y_a(0, 1, 0) + z_a(0, 0, 1)$$

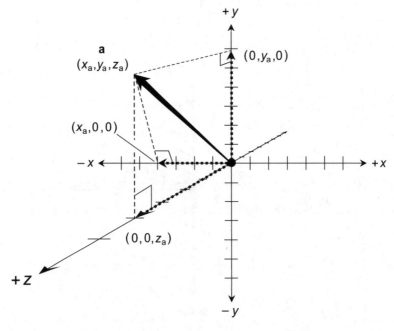

$$(x_a, y_a, z_a) = (x_a, 0, 0) + (0, y_a, 0) + (0, 0, z_a)$$

FIGURE 9-9 · In Cartesian three-space, we can break up any vector into a sum of three component vectors, each of which lies on one of the coordinate axes.

The vectors $(1,0,0)$, $(0,1,0)$, and $(0,0,1)$ are called *unit vectors* because their lengths all equal 1. Mathematicians and engineers name these three unit vectors **i**, **j**, and **k**, as follows:

$$(1,0,0) = \mathbf{i}$$
$$(0,1,0) = \mathbf{j}$$
$$(0,0,1) = \mathbf{k}$$

The vector **a** shown in Fig. 9-9 breaks down as

$$\mathbf{a} = (x_a, y_a, z_a)$$
$$= x_a\mathbf{i} + y_a\mathbf{j} + z_a\mathbf{k}$$

PROBLEM 9-6

Convert the vector $\mathbf{b} = (-2,3,-7)$ to a sum of multiples of the unit vectors **i**, **j**, and **k**.

SOLUTION

Envisioning the situation might require a keen "mind's eye," but you don't have to see the vectors to solve this problem. The vector breaks down neatly as

$$\mathbf{b} = (-2,3,-7)$$
$$= -2 \times (1,0,0) + 3 \times (0,1,0) + [-7 \times (0,0,1)]$$
$$= -2\mathbf{i} + 3\mathbf{j} + (-7)\mathbf{k}$$
$$= -2\mathbf{i} + 3\mathbf{j} - 7\mathbf{k}$$

Planes

The equation of a flat geometric plane in Cartesian three-space resembles that of a straight line in the Cartesian plane. We have an extra variable to contend with, but the general equation format is basically the same.

Criteria for Uniqueness

We can uniquely define a geometric plane in Euclidean three-space according to any of the following criteria:

- A point in the plane and a vector perpendicular to the plane
- Three points that don't all lie on the same straight line

- Two intersecting straight lines
- Two parallel straight lines

General Equation of Plane

The simplest possible equation for a plane in Cartesian three-space derives from the first of the foregoing criteria: a point in the plane and a vector that runs normal (perpendicular) to the plane. Figure 9-10 shows a plane W in Cartesian three-space, a point $P = (x_0, y_0, z_0)$ in plane W, and a vector $(a,b,c) = a\mathbf{i} + b\mathbf{j} + c\mathbf{k}$ oriented normal to plane W. In this example, the vector (a,b,c) originates at point P, not at the coordinate origin $(0,0,0)$, because W doesn't pass through the coordinate origin at all! Nevertheless, we can base the values $x = a$, $y = b$, and $z = c$ for the vector on the standard form, as if the vector did indeed start at the coordinate origin. Remember, all vectors having the same length and the same direction are in effect equal to one another, regardless of where their back-end (starting) points lie.

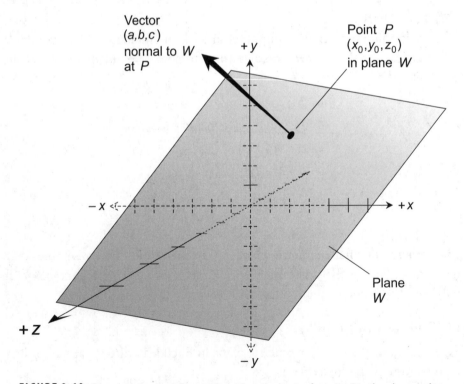

FIGURE 9-10 · We can uniquely define a plane W on the basis of a point P in the plane and a vector (a,b,c) normal to the plane. In this illustration, dashed portions of the coordinate axes lie "behind" the plane.

Once we know all these facts about a plane in Cartesian three-space, we have enough information to uniquely define that plane and write its equation as

$$a(x - x_0) + b(y - y_0) + c(z - z_0) = 0$$

We call the constants a, b, and c the *coefficients*. We can also write the equation as

$$ax + by + cz + d = 0$$

In this scenario, the value of d works out as

$$d = -(ax_0 + by_0 + cz_0)$$
$$= -ax_0 - by_0 - cz_0$$

Plotting a Plane

We can usually draw a graph of a plane in Cartesian three-space by determining the points where the plane crosses each of the three coordinate axes. We can then visualize the plane based on these points. Unfortunately, not all planes cross all three of the axes in Cartesian xyz-space. If a plane runs parallel to one of the axes, then that plane does not cross that axis. If a plane runs parallel to the plane formed by two of the axes, then that plane crosses only the axis to which it *does not* run parallel.

Still Struggling

Any plane in Cartesian three-space *must* cross *at least one* of the coordinate axes somewhere. Geometric planes have theoretically infinite extent. If we start at any point on a plane and travel around within that plane long enough, and if we venture far enough away from our starting point, eventually we'll "hit" at least one coordinate axis. In most planes, we'll eventually encounter two or all three of the coordinate axes.

PROBLEM 9-7

Draw a graph of the plane W represented by the equation

$$-2x - 4y + 3z - 12 = 0$$

✔ SOLUTION

To solve this problem, let's see if we can find the points where the plane crosses each of the coordinate axes. If we can find three such points, then we can use those points to define the plane. (If we can't find three such points, we'll have to try some other scheme, but let's not worry about that conundrum unless it comes up!)

We can find the *x*-intercept, or the point where the plane *W* intersects the *x* axis, by setting $y = 0$ and $z = 0$ and then solving for *x* as follows:

$$-2x - 4 \times 0 + 3 \times 0 - 12 = 0$$

Eliminating the "zero factors," we get

$$-2x - 12 = 0$$

When we add 12 to each side, we obtain

$$-2x = 12$$

Finally, we divide through by −2, getting

$$x = 12/(-2)$$
$$= -6$$

If we call the *x*-intercept point *P*, then

$$P = (-6,0,0)$$

We can find the *y*-intercept, or the point where the plane *W* intersects the *y* axis, by setting $x = 0$ and $z = 0$ in the original equation and then solving for *y* as follows:

$$-2 \times 0 - 4y + 3 \times 0 - 12 = 0$$

Eliminating the "zero factors" gives us

$$-4y - 12 = 0$$

Adding 12 to each side, we get

$$-4y = 12$$

Dividing through by −4 gives us the solution as

$$y = 12/(-4)$$
$$= -3$$

If we call the y-intercept point Q, then

$$Q = (0,-3,0)$$

We can determine the z-intercept, or the point where the plane W intersects the z axis, by setting $x = 0$ and $y = 0$ in the original equation and then solving for z as follows:

$$-2 \times 0 - 4 \times 0 + 3z - 12 = 0$$

When we get rid of the "zero factors," we have

$$3z - 12 = 0$$

We can add 12 to each side to obtain

$$3z = 12$$

Finally, we can derive the solution when we divide through by 3 to get

$$z = 12/3$$
$$= 4$$

If we call the z-intercept point R, then

$$R = (0,0,4)$$

The plot of Fig. 9-11 shows the three points P, Q, and R as we've derived them here:

$$P = (-6,0,0)$$
$$Q = (0,-3,0)$$
$$R = (0,0,4)$$

We can envision the position and orientation of the plane W on the basis of these three points, because they don't all lie along a single line.

Each division equals 1 unit

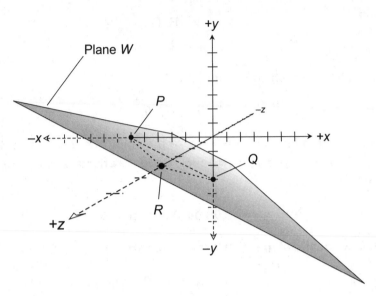

FIGURE 9-11 · Illustration for Problem 9-7. Dashed portions of the coordinate axes lie "behind" the plane.

PROBLEM **9-8**

Suppose that a plane contains the point (2,–7,0), and the vector $3i + 3j + 2k$ runs normal to the plane. What's the equation of the plane?

SOLUTION

The vector $3i + 3j + 2k$ is equivalent to $(a,b,c) = (3,3,2)$. We know the coordinates of one point in the plane; they are $(x_0, y_0, z_0) = (2,-7,0)$. Recall the general formula for the equation of a plane in Cartesian three-space:

$$a(x - x_0) + b(y - y_0) + c(z - z_0) = 0$$

Plugging our known values $a = 3$, $b = 3$, $c = 2$, $x_0 = 2$, $y_0 = -7$, and $z_0 = 0$ into this formula, we get

$$3(x - 2) + 3[y - (-7)] + 2(z - 0) = 0$$

Simplifying, we obtain

$$3(x - 2) + 3(y + 7) + 2z = 0$$

When we multiply out the terms in full, we have

$$3x - 6 + 3y + 21 + 2z = 0$$

which streamlines to

$$3x + 3y + 2z = -15$$

Straight Lines

Straight lines in Cartesian three-space present a more complicated picture than those in the Cartesian coordinate plane because we have an added dimension, making the expression of the direction more complicated. But all linear equations, no matter what the number of dimensions, have one thing in common: We can reduce any such equation to a form where no variable is raised to any power other than 0 or 1.

Symmetric-Form Equation

We can represent a straight line in Cartesian three-space using a "three-way" equation in three variables. Mathematicians call this type of expression a *symmetric-form equation*. It takes the following form, where x, y, and z represent the variables, (x_0, y_0, z_0) represents the coordinates of a specific point on the line, and a, b, and c represent real-number constants:

$$(x - x_0)/a = (y - y_0)/b = (z - z_0)/c$$

If we want this equation to make sense, none of the three constants a, b, or c can equal zero. If $a = 0$ or $b = 0$ or $c = 0$, then we end up with a zero denominator in one of the expressions, making it meaningless.

Direction Numbers

In the symmetric-form equation of a straight line, the constants a, b, and c are known as the *direction numbers* for that line. If we consider a vector **m** with its back-end point at the origin and its "arrowed end" at the point $(x,y,z) = (a,b,c)$, then the vector **m** runs parallel to the line denoted by the symmetric-form equation. We have

$$\mathbf{m} = a\mathbf{i} + b\mathbf{j} + c\mathbf{k}$$

where **m** constitutes the 3D equivalent of the slope of a line in the 2D Cartesian plane. Figure 9-12 illustrates a situation of this sort for a line L containing a point $P = (x_0, y_0, z_0)$ in Cartesian three-space.

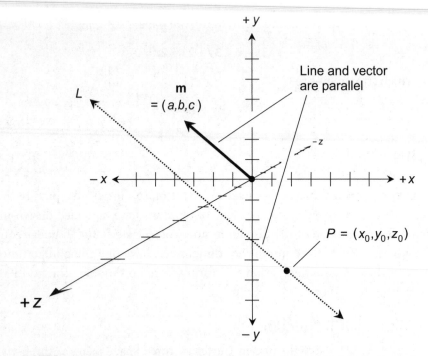

FIGURE 9-12 · We can uniquely define a line L on the basis of a point P on the line and a vector $\mathbf{m} = (a,b,c)$ that runs parallel to the line.

Parametric Equations

As you might suspect, infinitely many vectors can satisfy the requirement for $\mathbf{m}$ as we defined it earlier. If we let t represent any nonzero real number, then the vector

$$\mathbf{tm} = (ta,tb,tc)$$
$$= ta\mathbf{i} + tb\mathbf{j} + tc\mathbf{k}$$

will work every bit as well as $\mathbf{m}$ for the purpose of defining the direction of a line L. This handy fact leads to an alternative form for the equation of a line in Cartesian three-space in the form of three equations:

$$x = x_0 + at$$
$$y = y_0 + bt$$
$$z = z_0 + ct$$

We call the nonzero real number t a *parameter*, and the above three expressions a set of *parametric equations* for a straight line in Cartesian *xyz*-space.

TIP *If we want to define an entire geometric line (perfectly straight and infinitely long) on the basis of parametric equations, we must allow the parameter t to range over the entire set of real numbers, including zero.*

TIP *If any of the direction numbers for a line in Cartesian three-space happens to equal 0, then we must use parametric equations to describe the line. We can't use the symmetric form because it produces a denominator of 0 in one of the symmetric expressions.*

 **PROBLEM 9-9**

Find the symmetric-form equation for the line *L* shown in Fig. 9-13. Assume that the vector **m**, as shown, runs parallel to *L*.

SOLUTION

The figure shows us that line *L* passes through the point

$$P = (-5, -4, 3)$$

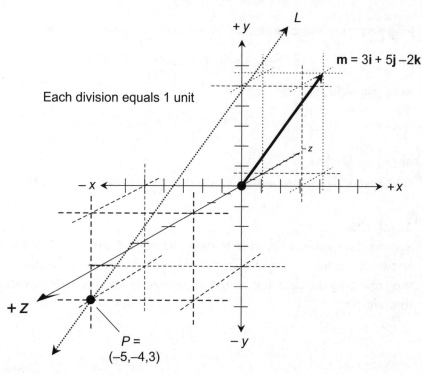

FIGURE 9-13 · Illustration for Problems 9-9 and 9-10.

We've been assured that line *L* runs parallel to the vector

$$m = 3i + 5j - 2k$$

The direction numbers for *L* equal the coefficients of m, as follows:

$$a = 3$$
$$b = 5$$
$$c = -2$$

We're given a specific point *P* on line *L*. If we say that $P = (x_0, y_0, z_0)$, then

$$x_0 = -5$$
$$y_0 = -4$$
$$z_0 = 3$$

We recall the general symmetric-form equation for a line in Cartesian three space as

$$(x - x_0)/a = (y - y_0)/b = (z - z_0)/c$$

Plugging the above coordinates for *P* into this equation, we get

$$[x - (-5)]/3 = [y - (-4)]/5 = (z - 3)/(-2)$$

which simplifies to

$$(x + 5)/3 = (y + 4)/5 = (z - 3)/(-2)$$

PROBLEM 9-10

Find a set of parametric equations for the line *L* shown in Fig. 9-13.

SOLUTION

Solving this problem involves merely rearranging the values of the six known values for x_0, y_0, z_0, *a*, *b*, and *c* in the symmetric-form equation, and then rewriting the data in the form of parametric equations. When we do that, we get

$$x = -5 + 3t$$
$$y = -4 + 5t$$
$$z = 3 - 2t$$

QUIZ

Refer to the text in this chapter if necessary. A good score is eight correct. Answers are in the back of the book. Use a calculator if you need one.

1. Imagine a unit circle in the Cartesian plane, and a ray that runs from the origin (0,0) outward and *downward toward the right* at an angle with respect to the *x* axis, so that we have to turn 60° *clockwise* to get from the *x* axis to the ray. What's the *x*-value of the point where the ray passes through the unit circle, accurate to three decimal places?
   A. 0.500
   B. −0.500
   C. 0.866
   D. −0.866

2. In the situation of Question 1, what's the *y*-value of the point where the ray passes through the unit circle?
   A. 0.500
   B. −0.500
   C. 0.866
   D. −0.866

3. Suppose that a vector in the Cartesian plane originates at the point (3,−7) and ends at the point (−7,3). What's the equivalent vector in standard form?
   A. (10,10)
   B. (−10,10)
   C. (10,−10)
   D. (−10,−10)

4. Figure 9-14 shows two vectors called a and b, both of which share a common back-end (originating) point, and which lie exactly perpendicular to each other. The cross-product vector a × b runs
   A. straight up.
   B. straight down.
   C. in the plane containing **a** and **b**, somewhere between them.
   D. nowhere, because it's the zero vector.

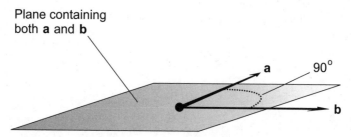

Plane containing both **a** and **b**

90°

a

b

FIGURE 9-14 · Illustration for Quiz Questions 4 through 6.

5. In the situation of Fig. 9-14, suppose that vector a measures exactly 1.2 units long while vector b measures exactly 0.8 unit long. What's the exact value of a • b?

   A. 0
   B. 0.96
   C. 1.5
   D. 2

6. In the situation of Fig. 9-14, the sum vector a + b runs

   A. straight up.
   B. straight down.
   C. in the plane containing **a** and **b**, somewhere between them.
   D. nowhere, because it's the zero vector.

7. What's the sum of the Cartesian xy-plane vectors (3,−7) and (6,2)?

   A. (9,−5)
   B. (−9,5)
   C. (−4,8)
   D. The sum vector does not lie in the Cartesian xy-plane.

8. What's the dot product of the Cartesian xy-plane vectors (3,−7) and (6,2)?

   A. 0
   B. 4
   C. −10
   D. −126

9. What's the dot product of the vectors 4i + 2j − 3k and −2i + 4j + 7k?

   A. −8i + 8j − 21k
   B. 14
   C. −21
   D. We need more information to answer this question.

10. In Cartesian three-space, the equation $3x − 4y − 17z = 10$ represents a

   A. straight line that passes through the origin.
   B. straight line that does not pass through the origin.
   C. plane that passes through the origin.
   D. plane that does not pass through the origin.

chapter 10

Alternative Coordinates

Cartesian coordinates aren't the only way that we can locate and define points in Euclidean two-space or three-space. Let's learn how some other coordinate systems work in two and three dimensions.

CHAPTER OBJECTIVES

In this chapter, you will

- Define two-space coordinates in terms of distance and direction.
- Examine simple geometric objects in polar coordinates.
- "Compress" an infinite coordinate plane into a finite region.
- Convert between Cartesian two-space and polar coordinates.
- Learn how to define spatial orientation in terms of latitude and longitude.
- See how astronomers and navigators define directions and locate points in the heavens.
- Define three-space coordinates in cylindrical and spherical terms.

Polar Coordinates

Figures 10-1 and 10-2 show two versions of the *polar coordinate plane*. We plot increasing values of the independent variable as angles θ going counterclockwise from a reference axis pointing to the right (or "east"). We plot increasing values of the dependent variable as a distance r (called the *radius*) going straight outward from the origin in any direction. Therefore, we can denote the coordinates of a point on the plane as an ordered pair (θ, r).

The Radius

In the polar coordinate plane, the radius increments appear as concentric circles. As the size of the circle increases, so does the value of r. In Figs. 10-1 and 10-2, we haven't labeled the concentric circles in radial units. You can do that to fit your own needs. Imagine each concentric circle, working outward, as increasing by any number of units that you want. For example, when you move from a given radial division (circle) outward to the next larger one, it might represent a radius increase of 1, 5, 10, or 100 units.

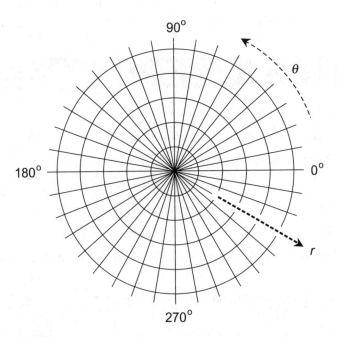

FIGURE 10-1 · The polar coordinate plane. In this rendition, we specify the angle θ in degrees and the radius r in uniform increments.

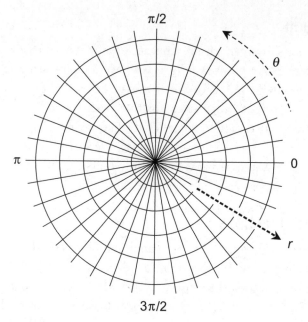

FIGURE 10-2 · Another form of the polar coordinate plane. In this case, we specify the angle θ in radians and the radius r in uniform increments.

TIP *No matter what increment "rate" you choose, the change in radius value between any two concentric circles should always equal the change in radius value between any two other adjacent concentric circles. In more technical terms, the radial axis should be linear.*

The Direction

We can express the direction in the polar plane in degrees or radians counterclockwise from a reference axis pointing to the right or "east." In Fig. 10-1, the direction θ appears in degrees. Figure 10-2 shows the same polar plane, using radians to express the direction. We don't need to use the "rad" abbreviation here; it's obvious that radians are intended from the fact that the angles all constitute multiples of π.

TIP *Regardless of whether you express angles in degrees or radians, you should make certain that the angular scale in the polar plane proceeds in a linear fashion. In other words, the physical angle on the graph should always vary in direct proportion to the value of the angle θ as you turn counterclockwise from the reference axis.*

Negative Radii

In polar coordinates, we can specify a negative value for radius and nevertheless define the position of a point as long as we have an expression for its angle as well. If we encounter a point for which r is given as a negative value, we can multiply r by –1 so that it becomes positive, and then add or subtract 180° (π rad) to or from the direction angle. That's like saying, "Travel 10 kilometers east" instead of "Travel –10 kilometers west." We must allow negative radius values in our polar system if we want to fully render graphs for mathematical functions whose ranges can attain negative values.

Nonstandard Directions

It's okay to have nonstandard direction angles in polar coordinates: angles that represent rotation through more than a full circle, or angles that represent clockwise rotation rather than counterclockwise rotation. If the value of θ equals 360° (2π rad) or more, it represents at least one, and likely more than one, complete counterclockwise rotation from the 0° (0 rad) reference axis. If the direction angle is less than 0° (0 rad), it represents clockwise rotation from the reference axis instead of counterclockwise rotation.

TIP   *We must allow nonstandard direction angles in order to graph figures that represent functions whose domains stray outside the standard span of angular values (i.e., outside the span $0° \leq \theta < 360°$ or $0 \leq \theta < 2\pi$).*

PROBLEM 10-1
Provide an example of a geometric object that represents a true mathematical function when we draw it on a polar coordinate plane, but not when we draw it on a Cartesian coordinate plane.

SOLUTION
Let's recall the definitions of the terms *relation* and *function* from Chap. 6. When we talk about a function f in polar coordinates, we can write $r = f(\theta)$. A simple function of θ in polar coordinates is a *constant function* such as

$$f(\theta) = 3$$

In polar coordinates, $f(\theta)$ constitutes an alternative way to denote r, the radius. Therefore, the above-defined function f tells us that $r = 3$. When we graph it in polar coordinates, we obtain a circle with a radius of 3 units, centered at the origin.

In Cartesian coordinates, the equation of the circle with radius of 3 units is more complicated than the polar equation. We have

$$x^2 + y^2 = 9$$

where $9 = 3^2$, the square of the radius. If we let y represent the dependent variable and x represent the independent variable in this situation, we can rearrange the equation of the circle to get

$$y = \pm (9 - x^2)^{1/2}$$

The circle having a radius of 3 units, and centered at the origin, represents a true mathematical function in polar coordinates, but not in Cartesian coordinates.

Still Struggling

If we say that $y = g(x)$ and then go on to claim that g constitutes a Cartesian coordinate function of x in the foregoing case, we're mistaken. There exist values of x (the independent variable) that produce two values of y (the dependent variable). For example, when we set $x = 0$, we end up with $y = \pm 3$. If we want to say that g is a relation in Cartesian coordinates, that's okay, but we can't call it a true mathematical function.

Some Examples

Circles, ellipses, spirals, and other figures with complicated equations in Cartesian coordinates can sometimes be portrayed more simply in polar coordinates. In the following examples, let's express the polar direction θ in radians by default.

Circle Centered at Origin

We can portray the equation of a *circle* centered at the origin in the polar coordinate plane with the general formula

$$r = a$$

where a represents a positive real-number constant. Figure 10-3 illustrates this situation.

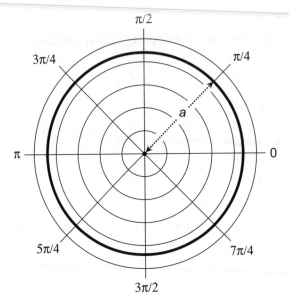

FIGURE 10-3 • Polar graph of a circle centered at the origin.

Circle Passing through Origin

The general form for the equation of a circle passing through the origin and centered at the point (θ_0, r_0) in the polar plane (Fig. 10-4) is

$$r = 2r_0 \cos (\theta - \theta_0)$$

Remember that the abbreviation "cos" refers to the trigonometric cosine function. In this case, we must find the cosine of the difference between two angles.

Ellipse Centered at Origin

We can determine the equation of an *ellipse* centered at the origin in the polar plane using the formula

$$r = ab/(a^2 \sin^2 \theta + b^2 \cos^2 \theta)^{1/2}$$

where a and b are positive real-number constants. Remember that the abbreviation "sin²" means the square of the sine of the indicated angle, while the abbreviation "cos²" means the square of the cosine of the indicated angle.

On an ellipse that we express in the foregoing manner, the constant a represents the distance from the origin to the curve as measured along the "horizontal" ray $\theta = 0$, and the constant b represents the distance from the origin to the

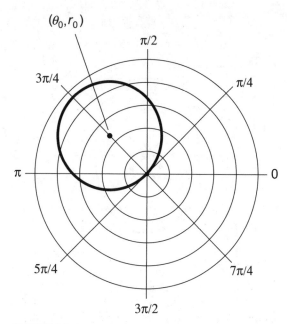

FIGURE 10-4 · Polar graph of a circle passing through the origin.

curve as measured along the "vertical" ray $\theta = \pi/2$. Figure 10-5 illustrates this arrangement.

TIP *The values a and b represent the lengths of the* semiaxes *of the ellipse. We call the greater of these two values the length of the* major semiaxis. *We call the lesser of these two values the length of the* minor semiaxis.

Hyperbola Centered at Origin

The general form of the equation of a *hyperbola* centered at the origin in the polar plane is

$$r = ab/(a^2 \sin^2 \theta - b^2 \cos^2 \theta)^{1/2}$$

where a and b represent positive real-number constants. This equation closely resembles the equation for the ellipse. However, instead of having a plus sign in the denominator, the hyperbola's equation has a minus sign there.

Imagine a rectangle D whose center lies at the coordinate origin, whose vertical edges lie tangent to the hyperbola, and whose vertices (corners) lie on the *asymptotes* of the hyperbola as shown in Fig. 10-6. (The asymptotes are the dashed lines

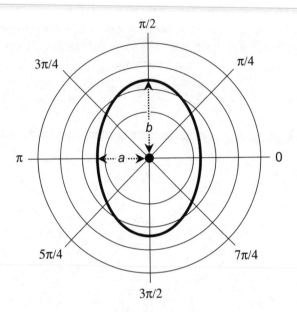

FIGURE 10-5 · Polar graph of an ellipse centered at the origin.

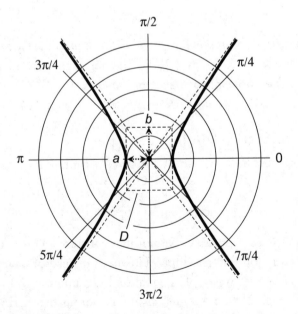

FIGURE 10-6 · Polar graph of a hyperbola centered at the origin.

that, combined, form a large "X" centered at the origin. The curves approach the asymptote lines as we move outward along them, but the curves never actually reach the asymptotes.) In the above general equation, a represents the distance from the origin to either vertical side of rectangle D as measured along the "horizontal" ray $\theta = 0$, and b represents the distance from the origin to either horizontal side of rectangle D as measured along the "vertical" ray $\theta = \pi/2$.

TIP   *The values a and b represent the lengths of the semiaxes of the hyperbola. We call greater of these two values the length of the major semiaxis. We call the lesser of these two values the length of the minor semiaxis.*

Lemniscate

The general form of the equation of a *lemniscate* (also called a *figure-eight*) centered at the origin in the polar plane is

$$r = a \, (\cos 2\theta)^{1/2}$$

where a represents a positive real-number constant. Figure 10-7 shows a generic situation of this sort. We can calculate the interior area A of each loop of the figure using the formula

$$A = a^2$$

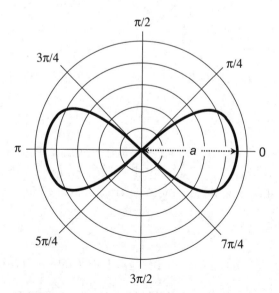

FIGURE 10-7 · Polar graph of a lemniscate centered at the origin.

Three-Leafed Rose

We can express the general form of the equation for a *three-leafed rose* centered at the origin in the polar plane as

$$r = a \cos 3\theta$$

or as

$$r = a \sin 3\theta$$

where a represents a positive a real-number constant. Figure 10-8A shows a generic cosine version of the curve. Figure 10-8B shows a generic sine version.

Four-Leafed Rose

The general form of the equation of a *four-leafed rose* centered at the origin in the polar plane is given by either of the following two formulas:

$$r = a \cos 2\theta$$

or

$$r = a \sin 2\theta$$

where a represents a positive a real-number constant. Figure 10-9A shows a generic cosine version of the curve. Figure 10-9B shows a generic sine version.

Spiral

The general form of the equation of a *spiral* centered at the origin in the polar plane is

$$r = a\theta$$

where a represents a positive a real-number constant. Figure 10-10 shows a generic example of this type of spiral, called the *spiral of Archimedes* because of the uniform manner in which its radius increases as the angle increases.

Cardioid

The general form of the equation for a *cardioid* centered at the origin in the polar plane is

$$r = 2a \left(1 + \cos \theta\right)$$

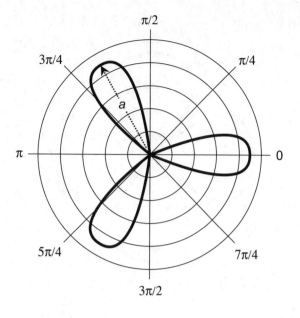

A

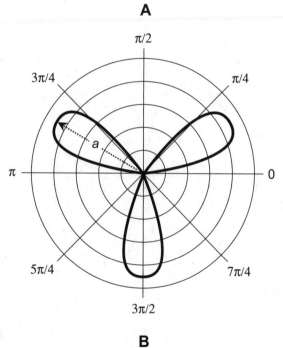

B

FIGURE 10-8 · A. Polar graph of a three-leafed rose with equation $r = a \cos 3\theta$. **B.** Polar graph of a three-leafed rose with equation $r = a \sin 3\theta$.

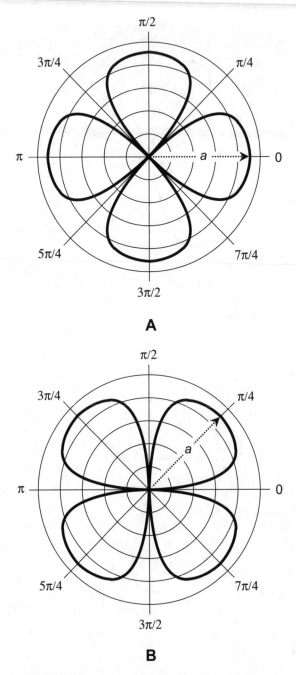

A

B

FIGURE 10-9 · A. Polar graph of a four-leafed rose with
equation $r = a \cos 2\theta$. **B.** Polar graph of a four-leafed rose
with equation $r = a \sin 2\theta$.

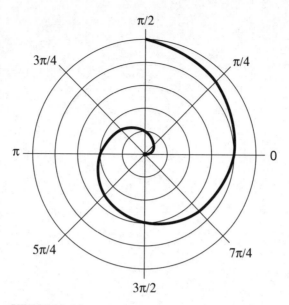

FIGURE 10-10 · Polar graph of a spiral; illustration for Problem 10-2.

where *a* represents a positive real-number constant. Figure 10-11 provides a generic example of this type of curve, also informally called a "heart" or "valentine."

 PROBLEM 10-2

If we let each radial division in Fig. 10-10 represent 1 unit, what's the equation of the spiral as shown?

 SOLUTION

Let's follow the curve outward, starting at the origin and proceeding counterclockwise, and then look at the radii for several specific angles:

- When $\theta = \pi/2$, we have $r = 1$
- When $\theta = \pi$, we have $r = 2$
- When $\theta = 3\pi/2$, we have $r = 3$
- When $\theta = 2\pi$, we have $r = 4$
- When $\theta = 5\pi/2$, we have $r = 5$

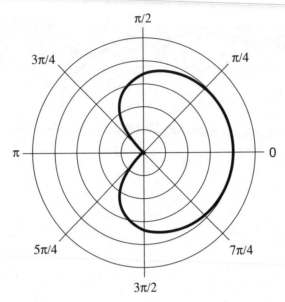

FIGURE 10-11 · Polar graph of a cardioid; illustration for Problem 10-3.

We can solve for *a* by substituting the foregoing number pairs in the general equation for the spiral. Actually, one point will suffice! We know that the point $(\theta, r) = (\pi, 2)$ lies on the spiral, and that's all we need. We have

$$r = a\theta$$

Substituting 2 in place of *r* and π in place θ, we obtain

$$2 = a\pi$$

When we divide through by π, we get

$$2/\pi = a$$

Now we know that $a = 2/\pi$, so the equation of the spiral must be

$$r = (2/\pi)\theta$$

or, without parentheses,

$$r = 2\theta/\pi$$

PROBLEM **10-3**

If we let each radial division in Fig. 10-11 represent 1 unit, what's the equation for the cardioid as shown?

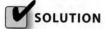

SOLUTION

Note that if $\theta = 0$, then $r = 4$. Let's solve for the constant in the general equation by "plugging" this number pair into that equation. Once again, the general equation is

$$r = 2a\,(1 + \cos\,\theta)$$

We know that $(\theta,r) = (0,4)$, so we have

$$4 = 2a\,(1 + \cos\,0)$$

Because $\cos 0 = 1$, we can simplify the above equation to

$$4 = 2a\,(1 + 1)$$

and further to

$$4 = 4a$$

Dividing through by 4, we get

$$1 = a$$

Now we know that $a = 1$, so the equation of the cardioid of Fig. 10-11 is

$$r = 2\,(1 + \cos\,\theta)$$

or, without parentheses,

$$r = 2 + 2\,\cos\,\theta$$

Compression and Conversion

Let's briefly examine a nonstandard coordinate system that can (at least) stimulate the imagination. Then we'll learn how to convert coordinate values between the polar plane and the Cartesian plane.

Geometric Polar Plane

Figure 10-12 shows a polar plane with a peculiar nonlinear radial scale: It's graduated geometrically instead of arithmetically (the usual case). The point

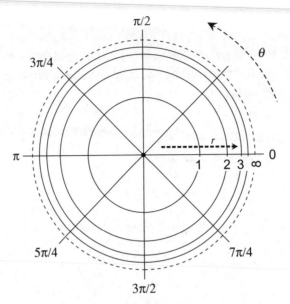

FIGURE 10-12 · A polar coordinate plane with a "geometrically compressed" radial axis.

corresponding to 1 on the *r* axis lies halfway between the origin and the outer periphery, which we label as ∞ (the "infinity" symbol). We place the points for radii of 2, 3, 4, and so on halfway between previous positive integer points and the outer periphery. In this way, we can portray the entire polar coordinate plane within an open circle of finite radius. The dashed circle at the outer extreme tells us that we do not actually define the value $r = \infty$.

We can expand or compress the radial scale of our *infinite polar coordinate* (IPC) system if we multiply or divide all the values on the *r* axis by a constant. This sort of modification allows us to plot a wide variety of relations and functions, minimizing distortion in particular regions of interest. All versions of the IPC introduce distortion into graphs that we draw. We observe the greatest distortion (relative to the conventional polar coordinate plane) near the periphery, and we observe the least distortion near the origin.

When we create an IPC system, we can use the same angular scale as we do with the ordinary polar coordinate plane. In Fig. 10-12, these angles appear in radians.

TIP *The foregoing "geometric axis compression" scheme also works with the axes of rectangular coordinates in two or three dimensions. You'll rarely (if ever) encounter schemes such as these in common mathematical literature, but they can provide "views to infinity" that other coordinate systems cannot do.*

Coordinate Conversions

Now that we've had a "glimpse of infinity," let's return to ordinary polar coordinates. Figure 10-13 shows a point $P = (x_0, y_0) = (\theta_0, r_0)$ graphed on superimposed Cartesian and polar coordinate planes. If we know the Cartesian coordinates, we can convert to polar coordinates using the following formulas:

$$\theta_0 = \arctan (y_0/x_0) \text{ if } x_0 > 0$$

$$\theta_0 = 180° + \arctan (y_0/x_0) \text{ if } x_0 < 0 \text{ (for } \theta_0 \text{ in degrees)}$$

$$\theta_0 = \pi + \arctan (y_0/x_0) \text{ if } x_0 < 0 \text{ (for } \theta_0 \text{ in radians)}$$

$$r_0 = (x_0{}^2 + y_0{}^2)^{1/2}$$

We can't have $x_0 = 0$ because that value produces an undefined quotient. If a value of θ_0 thus determined happens to be negative, we can add 360° or 2π rad to get the "legitimate" value.

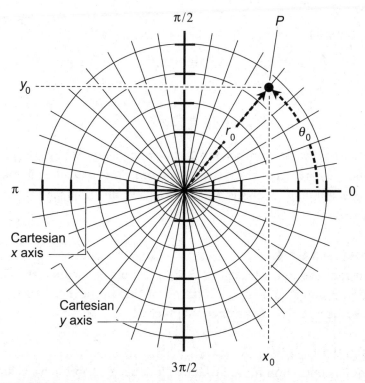

FIGURE 10-13 · Conversion between polar and Cartesian (rectangular) coordinates. Each radial division represents 1 unit. Each division on the x and y axes also represents 1 unit.

We can convert polar coordinates to Cartesian coordinates using the simpler formulas

$$x_0 = r_0 \cos \theta_0$$

and

$$y_0 = r_0 \sin \theta_0$$

Relation Conversions

We can use the foregoing formulas, in more generalized forms, to convert Cartesian-coordinate relations to polar-coordinate relations and vice versa. The generalized Cartesian-to-polar relation-conversion formulas appear as follows:

$$\theta = \arctan (y/x) \text{ if } x > 0$$
$$\theta = 180° + \arctan (y/x) \text{ if } x < 0 \text{ (for } \theta \text{ in degrees)}$$
$$\theta = \pi + \arctan (y/x) \text{ if } x < 0 \text{ (for } \theta \text{ in radians)}$$
$$r = (x^2 + y^2)^{1/2}$$

The generalized polar-to-Cartesian relation-conversion formulas are

$$x = r \cos \theta$$

and

$$y = r \sin \theta$$

TIP   *When you convert from polar to Cartesian coordinates or vice versa, a relation that's a function in one system might constitute a function in the other system as well—but not always. Make up a few examples and see what happens in each case.*

PROBLEM 10-4

Consider the point $(\theta_0, r_0) = (135°, 2)$ in polar coordinates. What's the ordered-pair (x_0, y_0) representation of this point in Cartesian coordinates, accurate to three decimal places?

SOLUTION

Use the conversion formulas above for specific coordinate values. Once again, they are

$$x_0 = r_0 \cos \theta_0$$

and

$$y_0 = r_0 \sin \theta_0$$

Plugging in the numbers produces the following values, rounded off to three decimal places:

$$x_0 = 2 \cos 135°$$
$$= 2 \times (-0.707)$$
$$= -1.414$$

and

$$y_0 = 2 \sin 135°$$
$$= 2 \times 0.707$$
$$= 1.414$$

Therefore, the coordinates of the point in the Cartesian plane are

$$(x_0, y_0) = (-1.414, 1.414)$$

The Navigator's Way

Navigators and military people use a form of polar coordinate plane similar to the one that mathematicians favor, except that the angle expression goes in the opposite direction. The radius is usually called the *range*, and real-world units are commonly specified, such as meters (m) or kilometers (km). The angle, or direction, is usually called the *azimuth*, *bearing*, or *heading*. We express this angle in degrees clockwise from north. Figure 10-14 shows the basic system. We symbolize the azimuth as α (the lowercase Greek alpha), and the range as r.

What Does "North" Mean?

At any point on the earth's surface, we have two distinct ways of defining "north," or 0°. The more accurate, preferred, and generally accepted standard uses *geographic north*, also known as *true north*. That's the direction in which we must travel over the surface if we want to follow the shortest possible route to the *north geographic pole*. The less accurate standard uses *magnetic north*, the direction indicated by the needle in a hiker's or mariner's magnetic compass.

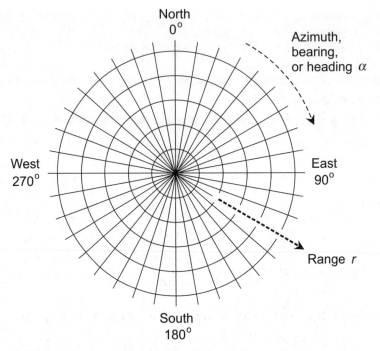

FIGURE 10-14 • The navigator's polar coordinate plane. We express the bearing α in degrees and the range r in real-world units.

Still Struggling

For most locations on the earth's surface, some difference exists between geographic north and magnetic north. This difference, measured in degrees, is called the *magnetic declination*, or sometimes simply the *declination*. Navigators in "the olden days" had to know the magnetic declination for their location whenever they couldn't use the stars to determine geographic north. Nowadays, most navigators have access to electronic navigation systems such as the *Global Positioning System* (GPS) that render the magnetic compass irrelevant—provided that all the hardware and software work properly! Even today, oceangoing vessels still have magnetic compasses on board in case of a failure of the more sophisticated equipment.

Strict Restrictions

In so-called *navigator's polar coordinates* (NPC), we can never have a negative value for the range. This constraint reflects the fact that in the "real world," nothing can lie any closer to us than the point where we stand! No navigator talks about traveling –20 kilometers on a heading of 270°, for example, when they really mean that they want to go 20 kilometers on a heading of 90°.

When we work out complicated navigational problems, we'll sometimes derive a negative value for the range. In a case of that sort, we should multiply the derived negative value of r by –1 (thereby making it positive with the same absolute value), and we should increase or decrease the value of α by 180° so that the azimuth remains at least 0° but less than 360°.

The azimuth, bearing, or heading in NPC must likewise conform to certain values. The smallest possible value of α is 0°, representing north. As we turn clockwise (as a bird might see it from some vantage point high above us), the values of α increase through 90° (east), 180° (south), 270° (west), and ultimately approach, but never reach, 360° (north again).

We can put the above restrictions into equation form quite simply: Whenever we use the NPC system of point location, we must have

$$0° \leq \alpha < 360°$$

and

$$r \geq 0$$

Mathematician's Polar versus Navigator's Polar

Once in awhile, we'll want to convert from *mathematician's polar coordinates* (MPC), which constitutes our default system, to NPC, or vice versa. The radius of a particular point, r_0, has exactly the same meaning in both systems, so that conversion process is trivial! However, the angle definitions differ between the two systems.

If we know the direction angle θ_0 of a point in MPC and we want to find the equivalent azimuth angle α_0 in NPC, we must make sure that we express θ_0 in degrees, not radians. Then, depending on the value of θ_0, we can use the conversion formulas

$$\alpha_0 = 90° - \theta_0 \text{ if } 0° \leq \theta_0 \leq 90°$$

or

$$\alpha_0 = 450° - \theta_0 \text{ if } 90° < \theta_0 < 360°$$

If we know the azimuth α_0 of a point in NPC and we want to find the equivalent direction angle θ_0 in MPC, then we can use one or the other of the following conversion formulas, depending on the value of α_0:

$$\theta_0 = 90° - \alpha_0 \text{ if } 0° \le \alpha_0 \le 90°$$

or

$$\theta_0 = 450° - \alpha_0 \text{ if } 90° < \alpha_0 < 360°$$

Navigator's Polar versus Cartesian

Suppose that we want to convert the position of a particular point from NPC to Cartesian coordinates. Here are the conversion formulas for translating the coordinates for a point (α_0, r_0) in NPC to a point (x_0, y_0) in the Cartesian xy-plane:

$$x_0 = r_0 \sin \alpha_0$$

and

$$y_0 = r_0 \cos \alpha_0$$

These formulas resemble the ones that we would use to convert MPC to Cartesian coordinates, except that the sine and cosine functions apply to different angles.

In order to convert the coordinates of a point (x_0, y_0) in Cartesian coordinates to a point (α_0, r_0) in NPC, we can use the following formulas:

$$\alpha_0 = \arctan (x_0/y_0) \text{ if } y_0 > 0$$
$$\alpha_0 = 180° + \arctan (x_0/y_0) \text{ if } y_0 < 0$$
$$r_0 = (x_0^2 + y_0^2)^{1/2}$$

Still Struggling

We can't have $y_0 = 0$ in the foregoing situation, because that would produce an undefined quotient. If a value of α_0 thereby determined turns out negative, we can add 360° to get the "legitimate" value.

PROBLEM 10-5

Imagine that a radar display uses NPC to indicate the presence of a hovering object at a bearing of 300° and a range of 40 kilometers. If we say that a kilometer equals a "unit" by default, what are the coordinates (θ_0, r_0) of this object in MPC? Express θ_0 in both degrees and radians.

SOLUTION

We know the NPC coordinates as $(\alpha_0, r_0) = (300°, 40)$. The value of r_0, the radius, equals the range, in this case 40 kilometers. As for the angle θ_0, we can recall the conversion formulas given above. In this case, because α_0 is greater than 90° and less than 360°, we have

$$\theta_0 = 450° - \alpha_0$$
$$= 450° - 300°$$
$$= 150°$$

It follows that

$$(\theta_0, r_0) = (150°, 40)$$

To express θ_0 in radians, recall that there are 2π rad in a full 360° circle or π rad in a 180° angle. Note that 150° equals exactly 5/6 of 180°. Therefore

$$\theta_0 = 5\pi/6 \text{ rad}$$

so we can say that

$$(\theta_0, r_0) = (150°, 40)$$
$$= (5\pi/6, 40)$$

TIP   *We can leave the "rad" off the angle designator in the foregoing situation. When we see no units specified for the measure of an angle, and if the figure contains some multiple or fraction of π, we can assume that radians are intended by default.*

PROBLEM 10-6

Imagine that you're traveling on an archeological expedition, and you unearth a stone tablet with a treasure map chiseled on its face. The map says "You are here" next to an X, and then says, "Go north 40 paces and then west 30 paces." Let the westerly compass direction correspond to the negative x axis of a Cartesian coordinate system. Let east correspond to the positive x axis, south correspond to the negative y axis, and north correspond to the positive y axis. Also suppose that you let one "pace" represent 1 "unit" of radius in the NPC system and also 1 "unit" on either axis in the Cartesian system. If you're ambitious enough to look for the treasure and lazy enough so you insist on walking in a straight line to reach it, how many paces should you travel, and in what direction, in NPC? Determine your answer to the nearest degree and to the nearest pace.

SOLUTION

Determine the ordered pair in Cartesian coordinates that corresponds to the imagined treasure site. Define the origin as the spot where you unearthed the map. If you let (x_0, y_0) represent the point on the earth's surface beneath which the treasure supposedly exists, then "40 paces north" translates to $y_0 = 40$, and "30 paces west" translates to $x_0 = -30$. Therefore

$$(x_0, y_0) = (-30, 40)$$

Because $y_0 > 0$, you can use the following formula to determine the heading α_0:

$$\alpha_0 = \arctan(x_0/y_0)$$
$$= \arctan(-30/40)$$
$$= \arctan -0.75$$
$$= -37°$$

To get this angle into the standard form, you must add 360°, obtaining

$$\alpha_0 = -37° + 360°$$
$$= 360° - 37°$$
$$= 323°$$

To find the value of the range r_0, you can use the distance formula

$$r_0 = (x_0^2 + y_0^2)^{1/2}$$

Plugging in $x_0 = -30$ and $y_0 = 40$, you get

$$r_0 = [(-30^2) + 40^2]^{1/2}$$
$$= (900 + 1600)^{1/2}$$
$$= 2500^{1/2}$$
$$= 50$$

You have now determined the NPC coordinates as

$$(\alpha_0, r_0) = (323°, 50)$$

You should walk 50 paces, approximately north by northwest. Then, if you have a shovel, you can go ahead and dig. Good luck!

Alternative 3D Coordinates

Let's look briefly at the basics of the most common coordinate systems that scientists, navigators, and mathematicians use when working on the surface of the earth or in "real-world" three-space.

Latitude and Longitude

We can use *latitude* and *longitude* angles to uniquely define the position of any point on the surface of a sphere. Figure 10-15A illustrates the system for defining or locating geographic points on the earth's surface. The *polar axis* connects two specified points that lie at *antipodes* (opposing surface locations) on the sphere. We assign these points latitude values of $\theta = 90°$ (north pole) and $\theta = -90°$ (south pole). The *equatorial axis* runs outward from the center of the sphere at a right angle to the polar axis. We assign this axis the longitude value of $\phi = 0°$.

We can express latitude θ in the positive sense (north) or the negative sense (south) with respect to the plane containing the equator. Longitude ϕ is measured counterclockwise (positively) and clockwise (negatively) relative to the equatorial axis. We restrict the ranges of the angle values to

$$-90° \le \theta \le 90°$$

for latitude, and

$$-180° < \phi \le 180°$$

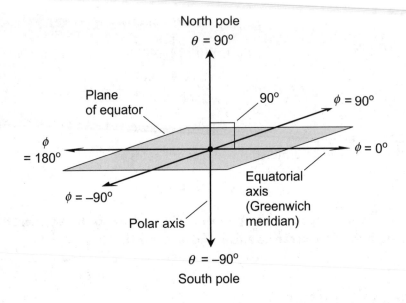

A

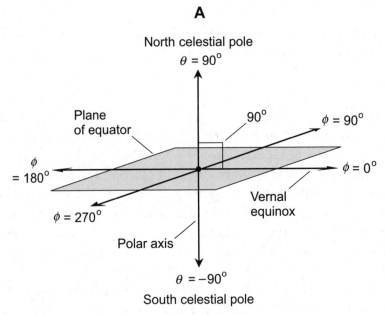

B

FIGURE 10-15 · A. Latitude and longitude coordinates for locating points on the earth's surface. **B.** Declination and right ascension coordinates for locating points in the sky.

for longitude. The latitude range includes the positive and negative extremes, but the longitude range includes only the positive extreme.

> **TIP** *On the earth's surface, the half-circle connecting the 0° longitude line with the poles passes through the town of Greenwich, England (not Greenwich Village in New York!) and is known as the* **Greenwich** *meridian or the* **prime** *meridian. Longitude angles are defined going east (positive) and west (negative) from the prime meridian.*

Celestial Coordinates

Celestial latitude and *celestial longitude* coordinates comprise extensions of the earth's latitude and longitude into the heavens. The set of coordinates that we use for geographic latitude and longitude applies to this system as well. An object with celestial latitude and longitude coordinates (θ,ϕ) appears at the *zenith* in the sky (directly overhead) from the point on the earth's surface with latitude and longitude coordinates (θ,ϕ).

Declination and *right ascension* define the positions of objects in the sky relative to the stars. Figure 10-15B portrays the essence of this system. The declination angle θ is identical to the celestial latitude. (Don't get *celestial declination* confused with *magnetic declination*, which we defined earlier in this chapter. The two parameters represent entirely different things!) We express the right ascension angle ϕ eastward along the *celestial equator* (a vast, imaginary circle in the sky, with the earth at its center, that lies in the same plane as the earth's equator) from the *vernal equinox* (the position of the sun in the heavens at the moment spring begins in the northern hemisphere, usually on March 20 or 21). We restrict the angle values to

$$-90° \leq \theta \leq 90°$$

and

$$0° \leq \phi < 360°$$

Note that the declination range includes the positive extreme and the negative extreme, but the right ascension range includes only the positive extreme.

Still Struggling

Astronomers sometimes use a specialized, rather peculiar scheme to define and measure the values of right ascension. Instead of expressing the angles in degrees or radians, they specify units of *hours*, *minutes*, and *seconds* based on 24 hours in a complete circle (corresponding to the 24 hours in a day). In that system, each hour of right ascension equals 15° (1/24 of a full circle). If that isn't confusing enough, minutes and seconds of right ascension differ from the chronological minutes and seconds that we encounter in everyday life, and also from the minutes and seconds of arc in the conventional geometric sense. One minute of right ascension equals 1/60 of an hour or 1/4 of a degree. One second of right ascension equals 1/60 of a minute or 1/240 of a degree. Nevertheless, in the case of declination angles, 1 minute equals 1/60 of an angular degree and 1 second equals 1/60 of a minute, or 1/3600 of a degree, the same as minutes and seconds of arc in the conventional geometric sense.

Cylindrical Coordinates

Figure 10-16 shows two systems of *cylindrical coordinates* for specifying the positions of points in three-space.

In the system of Fig. 10-16A, we start with Cartesian *xyz*-space. Then we define an angle θ in the *xy*-plane, in degrees or radians (but usually radians) turning counterclockwise from the positive *x* axis, which we call the *reference axis*. Given a point *P* in space, we consider its projection *P'* onto the *xy*-plane. We specify the position of *P* with the ordered triple (θ, r, h), defined as follows:

- The value of θ tells us the measure of the angle going counterclockwise from the reference axis to *P'* in the *xy*-plane.
- The value of *r* represents the distance or radius from the origin straight out to *P'*.
- The value of *h* represents the distance, called the *altitude* or *height*, of *P* above the *xy*-plane. (The point *P* lies below the *xy*-plane if and only if $h < 0$.)

Mathematicians, as well as some engineers and scientists, use this system of cylindrical coordinates.

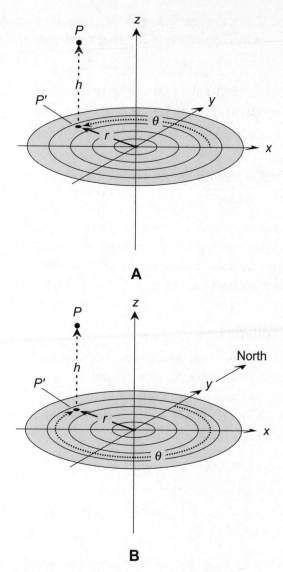

FIGURE 10-16 · A. Mathematician's cylindrical coordinates for defining points in three-space. **B.** Astronomer's and navigator's cylindrical coordinates for defining points in three-space.

In the system shown by Fig. 10-16B, we again start with Cartesian xyz-space. The xy-plane corresponds to the surface of the earth in the vicinity of the origin, and the z axis runs straight up (positive z values) and straight down (negative z values). The angle θ is defined in the xy-plane in degrees (but never radians) turning *clockwise* from the positive y axis, which corresponds to geographic north. Given a point P in space, we consider its projection P' onto the xy-plane. We specify the position of P with the ordered triple (θ, r, h), defined as follows:

- The value of θ tells us the measure of the angle going clockwise from geographic north to P' in the xy-plane.
- The value of r represents the distance (called the range) from the origin to P'.
- The value of h represents the altitude of P above the xy-plane. (The point P lies below the xy-plane if and only if $h < 0$.)

Navigators and aviators use this system of cylindrical coordinates to define or locate points in space over a limited region of the earth's surface. The system only works over a geographic region small enough so that the earth's curvature does not significantly affect the values.

Spherical Coordinates

Figure 10-17 shows three systems of *spherical coordinates* for defining points in space. The first two are used by astronomers and aerospace scientists, while the third one is of use to navigators and surveyors.

In the scheme of Fig. 10-17A, we specify the location of a point P with the ordered triple (θ, ϕ, r), defined as follows:

- The value of θ tells us the declination of P.
- The value of ϕ tells us the right ascension of P.
- The value of r tells us the distance (called the radius) from the origin to P.

In this example, we express the angles in degrees (except in the case of the astronomer's version of right ascension, which is expressed in hours, minutes, and seconds as defined earlier). Alternatively, we can express the angles in radians. This system remains fixed relative to the stars, even as the earth rotates.

Instead of declination and right ascension, the variables θ and ϕ can represent celestial latitude and celestial longitude, respectively, as shown in Fig. 10-17B. This system remains fixed relative to the earth. Therefore, the positions of celestial objects constantly change with time as the earth "turns underneath the heavens."

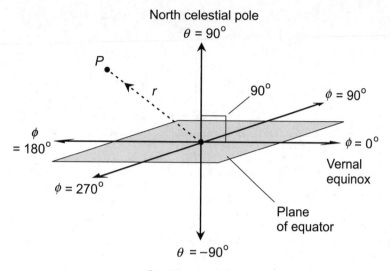

A

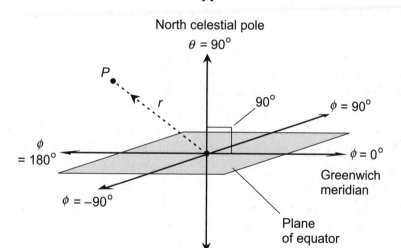

B

FIGURE 10-17 · A. Spherical coordinates for defining points in three-space, where the angles represent declination and right ascension. **B.** Spherical coordinates for defining points in three-space, where the angles represent celestial latitude and longitude.

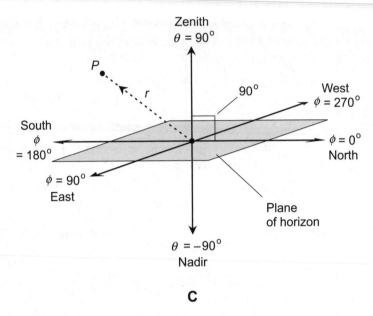

FIGURE 10-17 · C. Spherical coordinates for defining points in three-space, where the angles represent elevation (angle above the horizon) and azimuth (also called bearing or heading).

There's yet another alternative: θ can represent elevation (the angle above the horizon) and ϕ can represent the azimuth (bearing or heading), measured clockwise from geographic north for a specific location on the earth's surface. In this case, the reference plane corresponds to the horizon, not the equator, and the elevation can range between, and including, −90° (the nadir, or the point directly underfoot) and +90° (the zenith, or the point directly overhead). Figure 10-17C illustrates this system. Some people prefer to express the angle θ with respect to the zenith, rather than with respect to the horizon. In that case, the angular range becomes $0° \le \theta \le 180°$.

PROBLEM 10-7

What are the celestial latitude and longitude of the sun on the first day of spring in the northern hemisphere, when the sun lies at the vernal equinox in the plane of the earth's equator?

SOLUTION

The celestial latitude of the sun at the vernal equinox equals 0°, which equals the latitude of the earth's equator. The celestial longitude depends on the time of day. It's 0° (the Greenwich meridian) at high noon in

Greenwich, England or any other location at 0° longitude. From there, the celestial longitude of the sun proceeds west at the rate of 15° per hour of time (360° per 24-hour solar day).

PROBLEM 10-8

Imagine that you stand in a huge, perfectly flat field and fly a kite on a string 500 meters long. The wind blows directly from the east. The kite hovers at an altitude of 400 meters above the ground. If your body represents the coordinate origin and if you let the distance units of your system equal 1 meter, what's the position of the kite in the cylindrical coordinate scheme preferred by navigators and aviators?

SOLUTION

You can define the position of the kite with the ordered triple (θ, r, h), where θ represents the angle measured clockwise from geographic north to a point directly under the kite, r represents the distance from the origin to a point on the ground directly under the kite, and h represents the altitude of the kite above the ground. Because the wind blows from the east, you know that a point on the surface directly under the kite must lie west of the origin (represented by your body). Therefore $\theta = 270°$. The kite hovers at an altitude of 400 meters, so $h = 400$. You can find the value of r using the theorem of Pythagoras. You know that $h = 400$ units and the kite string measures 500 units in length, so

$$r^2 + 400^2 = 500^2$$

Expanding the squares of the numbers, you get

$$r^2 + 160{,}000 = 250{,}000$$

You can subtract 160,000 from each side to obtain

$$r^2 = 250{,}000 - 160{,}000$$

which simplifies to

$$r^2 = 90{,}000$$

and finally to

$$r = (90{,}000)^{1/2}$$

$$= 300$$

Therefore, in the system of cylindrical coordinates preferred by navigators and aviators, you can express the position of the kite in three-space as

$$(\theta, r, h) = (270°, 300, 400)$$

QUIZ

Refer to the text in this chapter if necessary. A good score is eight correct. Answers are in the back of the book.

1. Figure 10-18 is a polar-coordinate graph showing a particular point P. What's the x coordinate of P in the Cartesian xy-plane?

    A. -4

    B. -2

    C. $-8^{1/2}$

    D. $-\pi/2$

2. In the situation of Fig. 10-18, what's the y coordinate of point P in the Cartesian xy-plane?

    A. -4

    B. -2

    C. $-8^{1/2}$

    D. $-\pi/2$

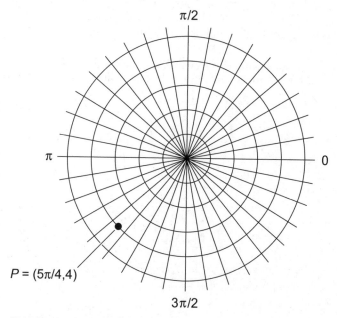

FIGURE 10-18 · Illustration for Quiz Questions 1 through 4.

3. In the graph of Fig. 10-18, suppose that we call the coordinate origin point Q. What's the equation of the open-ended ray QR in the polar coordinate system indicated here?

  A. $\theta = 5\pi/4$
  B. $\theta = r\pi/4$
  C. $r = -\pi/2$
  D. $r = 4\pi/\theta$

4. Suppose that you draw a Cartesian xy-plane coordinate grid directly on top of the polar coordinate grid in Fig. 10-18. Then you connect points Q and R with a straight line PQ that runs off forever in both directions. What's the equation of line PQ in the Cartesian xy-plane? Here's a hint: You'll need some of the knowledge that you gained in Chap. 6, along with what you learned in this chapter.

  A. $x = 4$
  B. $y = 5\pi/4$
  C. $y = 4x$
  D. $y = x$

5. Which of the following graphical objects portrays a true mathematical function of x in the Cartesian xy-plane but *does not* represent a true mathematical function of θ in MPC? Here's a hint: You'll need some of the knowledge that you gained in Chap. 6, along with what you learned in this chapter.

  A. A straight, vertical line that passes through the coordinate origin
  B. A straight, horizontal line that passes through the coordinate origin
  C. A circle that does not contain the coordinate origin
  D. A straight, horizontal line that does not pass through the coordinate origin

6. Figure 10-19 shows a point P in a cylindrical three-space coordinate system. The angle coordinate θ equals 140°, the radius coordinate r equals 6 units, and the

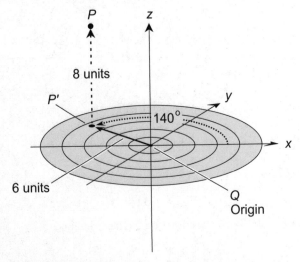

FIGURE 10-19 • Illustration for Quiz Questions 6 and 7.

altitude or height coordinate *h* equals 8 units as shown. We call the origin point *Q*. What's the length of line segment *PQ*, representing the direct distance in three-space between point *P* and the origin?

A. The square root of 48 units

B. 10 units

C. 12 units

D. 14 units

7. Suppose that in the situation of Question 6 and Fig. 10-19, we add 90° to the direction angle, thereby obtaining the coordinates $(\theta,r,h) = (230°,6,8)$ for point *P*. What happens to the length of line segment *PQ* in this case?

A. It becomes 230/140 as great.

B. It becomes 140/230 as great.

C. It does not change.

D. We need more information to answer this question.

8. One minute of right ascension, as an astronomer would define it, represents an angle equivalent to

A. 1/1440 of a full circle.

B. 1/720 of a full circle.

C. $1/(4\pi)$ of a full circle.

D. $1/\pi$ of a full circle.

9. Figure 10-20 shows a point *P* in a spherical three-space coordinate system where the angle θ represents celestial latitude, the angle ϕ represents right ascension (in degrees), and the radius *r* represents the distance from the origin, where you

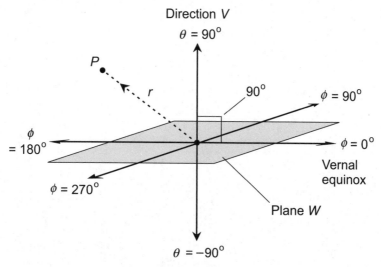

FIGURE 10-20 · Illustration for Quiz Questions 9 and 10.

stand as you observe the heavens. What does "Direction V" represent, as shown?

A. The zenith as you see it
B. The north geographic pole
C. The celestial equator
D. The north celestial pole

10. **In the situation of Question 9 and Fig. 10-20, what does W represent?**

A. The plane containing the earth's equator
B. The plane containing the earth's axis
C. The plane representing the horizon
D. The plane representing the earth's orbit around the sun

chapter **11**

Hyperspace and Warped Space

Some people can easily envision *hyperspace* (space of more than three dimensions) and *warped space*; others can't. Nevertheless, we can define them in geometric terms whether we can "see" them in our "mind's eyes" or not. Let's explore these esoteric concepts.

CHAPTER OBJECTIVES

In this chapter, you will

- Define Cartesian space of more than three dimensions (hyperspace).
- Learn how time-space "works."
- Quantify the relationship between time and distance.
- Envision simple four-dimensional objects.
- Calculate distances in hyperspace.
- Modify Euclid's fifth postulate.
- Take imaginary journeys in warped space.

Cartesian *n*-Space

As we have seen, the rectangular (or Cartesian) coordinate plane derives from two perpendicular number lines that intersect at their zero points. The lines form the coordinate axes, often called the *x* axis and the *y* axis. We can name or identify any point in this system as an ordered pair of the form (x,y). We call $(0,0)$ the origin. We can define Cartesian three-space using three number lines that intersect at a single point corresponding to the zero point of each line, and such that each line runs perpendicular to the plane determined by the other two lines. The lines form axes, representing variables such as x, y, and z. Points are defined by ordered triples of the form (x,y,z). The origin is $(0,0,0)$. Let's extrapolate the Cartesian-coordinate concept into more than three dimensions.

Four Spatial Dimensions

We can set up a system of rectangular coordinates in four dimensions—*Cartesian four-space* or *4D space*—using four number lines that intersect at a single point corresponding to the zero point of each line, and such that each of the lines runs perpendicular to the other three. The lines form axes, representing variables such as w, x, y, and z. Alternatively, we can label the axes x_1, x_2, x_3, and x_4. We can name or identify individual points as *ordered quadruples* of the form (w,x,y,z) or (x_1,x_2,x_3,x_4), defining the origin as the point represented by $(0,0,0,0)$.

> **TIP** *As with the variables or numbers in ordered pairs and triples, we never put any spaces after the commas when we write an ordered quadruple.*

At first you might think, "Cartesian four-space is easy to imagine," and draw a diagram such as Fig. 11-1 to illustrate it. But when we try to plot points in this system, we encounter a problem. We can't define points in this rendition of four-space without ambiguity. There aren't enough points in 3D space to pair off one-to-one with all possible values of the ordered quadruple (w,x,y,z). In three-space as we know it, we can't arrange four number lines such as those shown in Fig. 11-1 so that they intersect at a single point with each line running perpendicular to all three of the others.

Imagine one of the points in a room where two adjacent walls meet the floor. Unless the building has an unusual architecture or has begun to settle (sag) because of earth movement, this intersection defines three straight line segments.

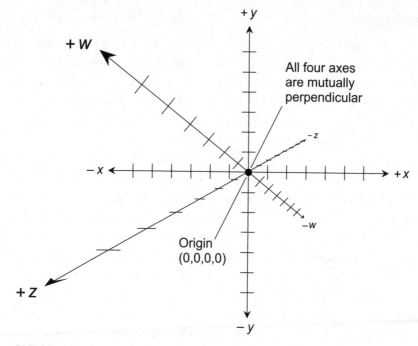

FIGURE 11-1 · Concept of Cartesian four-space. The *w*, *x*, *y*, and *z* axes are all mutually perpendicular at the origin point (0,0,0,0).

One of the segments runs up and down between the two walls, and the other two run horizontally between the two walls and the floor. The line segments are all mutually perpendicular at the point where they come together (where the walls meet the floor). The lines containing the three line segments can represent the *x*, *y*, and *z* axes in Cartesian three-space coordinate system.

Now try to envision a fourth line segment that has one end at the intersection point of the existing three line segments, and that runs perpendicular to them all. Such a line segment can't exist in ordinary space! But in four dimensions, or hyperspace, it can exist.

TIP *Mathematically, we can work with Cartesian four-space, even though most of us can't directly envision it. As things work out, we need four dimensions to completely describe points, objects, and events in the "real universe." Albert Einstein was one of the first scientists to put forth the idea that the "fourth dimension" exists in physical reality (as opposed to residing as abstract notions in mathematicians' minds).*

Time-Space

You've seen *time lines* in history books. You've seen them in graphs of quantities such as temperature, barometric pressure, or stock market prices plotted as functions of time. Isaac Newton, one of the most renowned mathematicians in the history of the Western world, imagined time as "flowing smoothly and unalterably." Time, according to so-called *classical physics* or *Newtonian physics*, does not depend on space, nor does space depend on time.

Wherever you are, however fast or slow you travel, and no matter what else you do, the "cosmic clock" (according to classical physics) keeps ticking at the same absolute rate. In everyday scenarios, this model works well; its imperfections are not evident to nonscientists. However, Newton's paradigm represents an oversimplification. It can't completely describe what really happens in the cosmos on a large scale, at high relative speeds, or in intense gravitational fields.

Let's imagine a time line passing through 3D space, "perpendicular" to all three spatial axes such as the intersections between two walls and the floor of a room. The time axis passes through three-space at some chosen origin point, such as the point where two walls meet the floor in a room, or the center of the earth, or the center of the sun, or the center of our galaxy.

In four-dimensional (4D) *Cartesian time-space* (or simply *time-space*), each point follows its own time line. Assuming that none of the points moves with respect to the origin, all the points follow time lines that run "parallel" to all the other time lines, and all the time lines run "perpendicular" to three-space. Figure 11-2 illustrates this concept in dimensionally reduced form (with one of the spatial dimensions taken away, so that three-space shows up as a Euclidean plane).

Position versus Motion

Imagine that we choose the sun as the origin point for a vast Cartesian three-space coordinate system. Suppose that the x and y axes lie in the plane of the earth's orbit around the sun. Also, suppose that the positive x axis runs from the sun through the earth's position in space on March 21 and thence onward into deep space (roughly toward the constellation Virgo). In this scenario

- The negative x axis runs from the sun through the earth's position on September 21 (roughly toward the constellation Pisces).
- The positive y axis runs from the sun through the earth's position on June 21 (roughly toward the constellation Sagittarius).

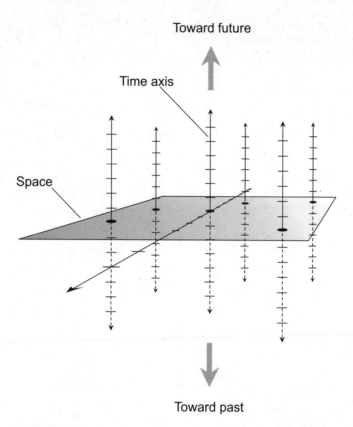

Toward future

Time axis

Space

Toward past

FIGURE 11-2 · Time as a fourth dimension. We illustrate three-space in dimensionally reduced form as a plane. Each stationary point in space follows a time line "perpendicular" to 3D space and "parallel" to the time axis.

- The negative y axis runs from the sun through the earth's position on December 21 (roughly toward the constellation Gemini).

- The positive z axis runs from the sun toward the north celestial pole (in the direction of Polaris, the North Star).

- The negative z axis runs from the sun toward the south celestial pole (where there's no prominent constellation).

Let's say that each division on the coordinate axes represent 1/4 of an *astronomical unit* (AU), where 1 AU equals the mean distance of the earth from the sun (about 150,000,000 kilometers). Figure 11-3A shows our new "deep-space coordinate" system, with the earth on the positive x axis, at a distance of 1 AU. The coordinates of the earth at this time are (1,0,0).

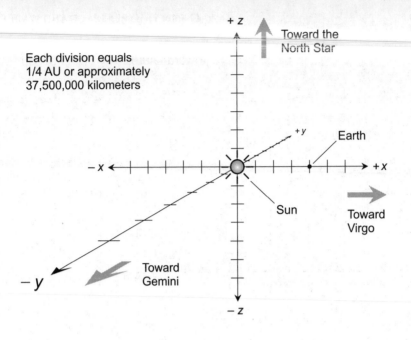

Each division equals
1/4 AU or approximately
37,500,000 kilometers

+z

Toward the
North Star

+y

Earth

−x

Sun

+x

Toward
Virgo

−y

Toward
Gemini

−z

A

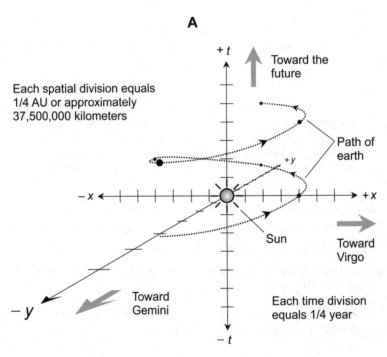

+t

Toward the
future

Each spatial division equals
1/4 AU or approximately
37,500,000 kilometers

Path of
earth

+y

−x

+x

Sun

Toward
Virgo

−y

Toward
Gemini

Each time division
equals 1/4 year

−t

B

FIGURE 11-3 · A. A Cartesian coordinate system for the position of the earth in 3D space. **B.** A dimensionally reduced Cartesian system for rendering the path of the earth through 4D time-space.

Of course, the earth doesn't remain fixed in space. It orbits the sun. Let's take away the z axis in Fig. 11-3A and replace it with a time axis called t. Now let's think hard: What does the earth's path look like in xyt-space if we let each increment on the t axis represent 1/4 of a year (90° of revolution around the sun)?

The earth's path through this dimensionally reduced time-space continuum does not constitute a straight line. Instead, when we follow the earth over time, we get a *helix* as shown in Fig. 11-3B. The earth's distance from the t axis remains nearly constant (it varies slightly because the earth's orbit around the sun does not form a perfect circle, but let's neglect that little detail). Every 1/4 of a year, the earth advances 90°, or one-quarter of a revolution, around the helix, and also moves forward by one increment along the time axis.

Some Hyper Objects

Now that we're no longer confined to 3D space, let's put our newly empowered imaginations to work. What characteristics do 4D objects and events have? How about objects and events in five dimensions (5D) and beyond?

Time as Displacement

When we consider time as a dimension, we need a standard—some sort of conversion factor—that relates time to spatial displacement. How many kilometers does 1 second of time comprise? At first, this question seems rather silly, akin to asking how many apples equal a gallon of water. But the more we ponder the notion, the more sensible it gets: We can relate time and displacement in terms of some known, constant *speed*.

Suppose that someone tells us, "The town of Jimsville is an hour away from the town of Joesville." We've all heard people talk like this, and we understand what they mean; the statement implies that we travel from one town to the other at a certain *rate of speed*. How fast must we drive a car to get from Jimsville to Joesville in an hour? If Jimsville and Joesville lie 50 kilometers from each other as measured along a stretch of highway, then we must travel at an average speed of 50 kilometers per hour in order to claim that Jimsville is an hour away from Joesville. If Jimsville lies 20 kilometers from Janesville, then we need only travel at an average speed of 20 kilometers per hour to say that Jimsville lies an hour away from Janesville.

Still Struggling

Do you remember the basic formula in classical physics that relates distance, speed, and time? In case you've forgotten, it's

$$d = st$$

where d represents the distance in kilometers that an object travels, s represents the object's speed in kilometers per hour, and t represents the number of hours that the object takes to traverse the specified distance. Using this formula, we can define time in terms of displacement and vice versa.

Universal Speed

The foregoing scheme allows us to convert time to distance in a relative way, depending on the speed at which we travel between two points. It's reasonable to ask, "Does any speed exist, some universal conversion factor, with which we can relate time and distance in an *absolute* sense?" According to Albert Einstein's *theory of special relativity*, the answer is a qualified "Yes."

The speed of light in a vacuum, commonly denoted c, remains constant regardless of the viewpoint (or *reference frame*) of any observer, as long as that observer does not accelerate at an extreme rate or sit in an extreme gravitational field. The constancy of c forms a fundamental principle of the theory of special relativity. The value of c lies close to 299,792 kilometers per second; let's round it off to 300,000 kilometers per second.

If d represents the distance between two points in kilometers and t represents the time in seconds that it takes for a ray of light to travel from one point to the other through empty space, then

$$d = ct$$
$$= 300,000\, t$$

According to this model, the moon, which orbits the earth at a distance of about 400,000 kilometers, is 1.33 *second-equivalents* distant from us. The sun is about 8.3 *minute-equivalents* away. The Milky Way galaxy is 100,000 *year-equivalents* in diameter. Astronomers call these units *light-seconds*, *light-minutes*, and *light-years*. We can also say that any two points in time separated by 1 second, but that occupy the same *xyz* coordinates in Cartesian three-space, lie 300,000 *kilometer-equivalents* apart as defined along the *t* axis.

At this moment yesterday, if you were in the same location as you now sit, your location in time-space was 24 (hours per day) times 60 (minutes per hour) times 60 (seconds per minute) times 300,000 (kilometers per second), or 25,920,000,000 kilometer-equivalents, away from where you are now.

Still Struggling

The above-described way of thinking takes quite a bit of getting-used-to! But after awhile, it starts to make a strange sort of sense. Consider this example: You might as well try to jump 25,920,000,000 kilometers in a single leap as try to change what happened in your own house 24 hours ago. You can no more alter history than you can fly through space like a light beam.

We can modify the foregoing conversion formula for smaller distances, more typical of everyday life. If *d* represents the distance in kilometers and *t* represents the time in milliseconds (units of 0.001 second), then

$$d = 300\,t$$

TIP *The above formula also holds for d in meters and t in microseconds (units of 0.000001, or 10^{-6}, second), and for d in millimeters (units of 0.001 meter) and t in nanoseconds (units of 0.000000001, or 10^{-9}, second), so we can speak of meter-equivalents, millimeter-equivalents, microsecond-equivalents, or nanosecond-equivalents.*

The Four-Cube

Imagine some of the simple, regular polyhedra in Cartesian four-space. What are their properties? Think about a *four-cube*, also known as a *tesseract*. This is an object with several identical 3D *hyperfaces*, all of which comprise cubes. How many vertices does a tesseract have? How many edges? How many 2D faces? How many 3D hyperfaces? How can we envision such an object to figure out the answers to these questions?

We can't make a 4D model of a tesseract out of toothpicks to examine its properties, and most people (if any) can't "see" a tesseract in their "mind's eyes" at all. But we can imagine a cube that appears from nothing, exists for awhile, and then disappears, such that it "lives" for a length of time equivalent to the length of any of its spatial edges and does not move during its existence. Because we've defined an absolute relation between time and displacement (the speed of light in a vacuum), we can graph a tesseract in which each edge has a length of, say, 300,000 kilometer-equivalents. This object is an ordinary 3D cube that measures 300,000 kilometers along each edge. It appears at a certain time t_0 and then disappears precisely 1 second later, at $t_0 + 1$. The sides of the cube each measure 1 second-equivalent in length, and the cube "lives" for 300,000 kilometer-equivalents of time.

Figure 11-4A shows a tesseract in dimensionally reduced form. Each division along the x and y axes represents 100,000 kilometers (the equivalent of 1/3 second), and each division along the t axis represents 1/3 second (the equivalent of 100,000 kilometers). Figure 11-4B portrays the tesseract in another way, as two 3D cubes (in perspective) connected by dashed lines representing the passage of time.

The Rectangular Four-Prism

A tesseract is a special form of the more general figure, known as a *rectangular four-prism* or *rectangular hyperprism*. Such an object consists of a 3D rectangular prism that abruptly comes into existence, lasts a certain length of time, disappears all at once, and does not move during its "lifetime." Figure 11-5 shows two examples of rectangular four-prisms in dimensionally reduced time-space.

Suppose the height, width, depth, and lifetime of a rectangular hyperprism, all measured in kilometer-equivalents, equal h, w, d, and t, respectively. We can calculate the *4D hypervolume* of this object (call it V_{4D}), in *quartic kilometer-equivalents*, as the product

$$V_{4D} = hwdt$$

The mathematics works in the same way if we express the height, width, depth, and lifetime of the object in second-equivalents. In that case, the 4D hypervolume equals the product $hwdt$ in *quartic second-equivalents*.

Impossible Paths

Certain paths are impossible to follow in Cartesian 4D time-space as we've defined it here. According to Einstein's special theory of relativity, nothing can

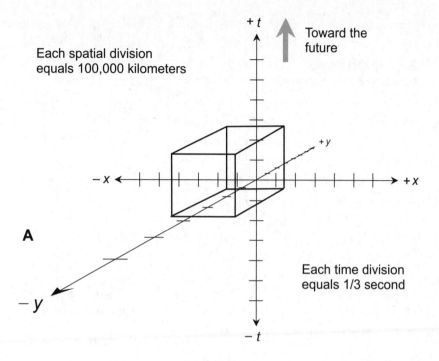

+t

Toward the
future

Each spatial division
equals 100,000 kilometers

+y

−x   +x

A

Each time division
equals 1/3 second

−y

−t

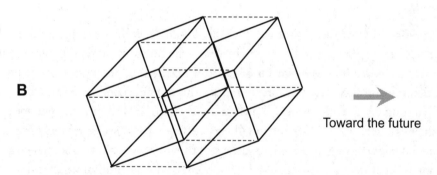

B

Toward the future

FIGURE 11-4 · At A, a dimensionally reduced plot of a time-space tesseract. At B,
another rendition of a tesseract, portraying time as lateral motion.

travel faster than the speed of light in *free space* (a vacuum). This physical law
restricts the directions in which line segments, lines, and rays can run when they
represent the trajectories of real objects in motion.

Consider what happens in 4D Cartesian time-space when we switch on a light
bulb. Imagine that the bulb rests at the origin of a Cartesian coordinate system,
and nothing but empty space surrounds the bulb for millions of kilometers in

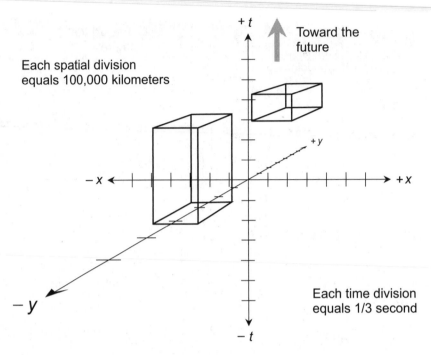

FIGURE 11-5 · Dimensionally reduced plots of two rectangular hyperprisms in time-space.

every direction. At the instant we close the switch, thereby powering up the light bulb, *photons* (particles of light) emerge from the bulb. In the first few moments, the initial, or leading, light travels outward from the bulb in expanding spherical paths or fronts. If we dimensionally reduce this situation and graph it, we get an expanding circle centered on the time axis, which, as time passes, generates a cone as shown in Fig. 11-6. In true 4D space, the actual geometric figure constitutes a *hypercone* or *four-cone*. The surface of the four-cone has two spatial dimensions (which portray the surface of a sphere) and one time dimension (which portrays the expansion of the sphere). Physicists call it a *light cone*.

Imagine an object that starts out at the location of the light bulb, and then moves away from the bulb immediately when we apply power to the bulb. In any real-life physical situation, the object must follow a path that remains entirely within the light cone defined by the initial photons from the bulb. Figure 11-6 shows one plausible path and one implausible path. If an object could travel outside the light cone, that object would move faster than the speed of light relative to the bulb—but that's impossible according to the theory of special relativity.

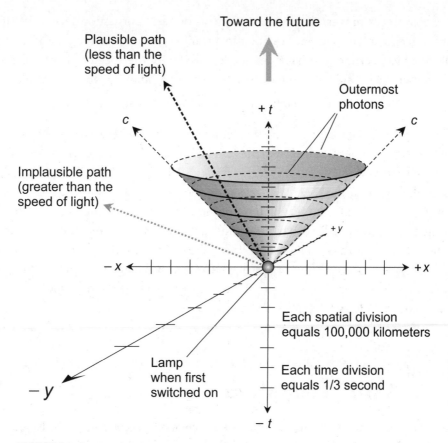

FIGURE 11-6 · Dimensionally reduced plot of the leading photons from a light bulb. Paths inside the cone represent relative speeds less than *c* (the speed of light); paths outside the cone represent relative speeds greater than *c*.

General Time-Space Hypervolume

Imagine an object—any object—in 3D space. Suppose that its spatial volume in cubic kilometer-equivalents equals a fixed quantity; let's call it V_{3D}. Suppose that such an object appears from nowhere, lasts a certain length of time *t* in kilometer-equivalents, and then ceases to exist. Further imagine that this object does not move with respect to us, the observers, at any time during its "lifetime." In this case, we can calculate the object's 4D time-space hypervolume V_{4D} using the formula

$$V_{4D} = V_{3D} \, t$$

The 4D time-space hypervolume of any object equals its spatial volume multiplied by its lifetime, provided that we express the time and displacement in equivalent units, and as long as the object never moves relative to us.

If an object moves, then we must incorporate a "correction factor" in the above formula. This factor does not affect things very much as long as the speed of the object (call it s) remains small compared with the speed of light c. But if s represents a considerable fraction of c, we must modify the above formula to

$$V_{4D} = V_{3D}t\,(1 - s^2/c^2)^{1/2}$$

The correction factor, $(1 - s^2/c^2)^{1/2}$, is close to 1 when s equals a small fraction of c and approaches 0 as s approaches c. This correction factor derives from the special theory of relativity.

In this context, the speed s always represents a relative quantity. It depends on the point of view from which an observer witnesses or measures it. If we want the term "speed" to have meaning, we must always add the qualifying phrase "relative to a certain observer." In these examples, we envision motion as taking place relative to the origin of a 3D Cartesian system, which translates into lines, line segments, or rays pitched at various angles with respect to the time axis in a 4D time-space Cartesian system.

Still Struggling

If you're still confused about kilometer-equivalents and second-equivalents, you can refer to Table 11-1 for reference. Keep in mind that time and displacement relate according to the equation

$$d = ct$$

where d represents the displacement (in linear units), t represents the time (in time units), and c represents the speed of light in linear units per unit time, as it travels through free space. Using this conversion formula, you can "morph" any displacement unit into an equivalent time interval and any time interval into an equivalent displacement unit.

PROBLEM 11-1

How many second-equivalents compose a distance of 1 kilometer?

SOLUTION

We know that the speed of light equals 300,000 kilometers per second (accurate to three significant figures) in free space, so it takes 1/300,000 of

TABLE 11-1 Displacement and time equivalents in free space (a vacuum) where the speed of light equals approximately 300,000 kilometers per second. Consider the displacement equivalents accurate to three significant figures.

Displacement Equivalent	Time Equivalent
9,460,000,000,000 kilometers	1 year
25,900,000,000 kilometers	1 solar day
1,079,000,000 kilometers	1 hour
18,000,000 kilometers	1 minute
300,000 kilometers	1 second
300 kilometers	0.001 second
1 kilometer	0.00000333 second
300 meters	0.000001 second
1 meter	0.00000000333 second
300 millimeters	0.000000001 second
1 millimeter	0.00000000000333 second

a second for a ray of light to travel 1 kilometer. That's approximately 0.00000333 second or 3.33 microseconds. One kilometer therefore represents 0.00000333 second-equivalent, or 3.33 microsecond-equivalents.

Beyond Four Dimensions

No limit exists as to the number of dimensions that we can define using the Cartesian coordinate paradigm. We can "create" spaces having any positive whole number of dimensions—10, 20, 100, 200, or whatever! We can incorporate time as a dimension if we want, but we don't *have* to include it.

Cartesian Extrapolations

A system of rectangular coordinates in five dimensions defines *Cartesian five-space*. This system has five number lines that serve as coordinate axes, all of which intersect at a point corresponding to the zero point of each line, and such that each of the lines runs perpendicular to the other four. We can call the variables for the resulting axes v, w, x, y, and z. Alternatively, we might call them x_1, x_2, x_3, x_4, and x_5. Points are named or identified by *ordered quintuples* such as (v,w,x,y,z) or (x_1,x_2,x_3,x_4,x_5). The origin point has the coordinates $(0,0,0,0,0)$. As you can guess, it doesn't matter what we call the variables, as long as we allow each one to change value independently from the other four.

A system of rectangular coordinates in *Cartesian n-space* (where n represents any positive integer, as large as we want) consists of n number lines, all of which intersect at their zero points, such that each of the lines runs perpendicular to all the others. The axes can be named x_1, x_2, x_3, ..., and so on up to x_n. Points in Cartesian n-space can be uniquely defined by ordered n-tuples of the form $(x_1,x_2,x_3,...,x_n)$.

A Five-Prism

Imagine a tesseract or a rectangular four-prism that appears at a certain time, does not move, and then disappears some time later. This object constitutes a *rectangular five-prism*. If x_1, x_2, x_3, and x_4 represent four spatial dimensions (in kilometer-equivalents or second-equivalents) for a rectangular four-prism in Cartesian four-space, and if t represents the five-prism's "lifetime" in the same units, then the *5D hypervolume* (call it V_{5D}) equals the product of them all. We have

$$V_{5D} = x_1 x_2 x_3 x_4 t$$

This formula holds only as long as the five-prism doesn't move relative to us at a significant speed. If the prism moves at a fast enough speed, then we must incorporate the relativistic correction factor $(1 - s^2/c^2)^{1/2}$, where s represents the object's relative speed.

Dimensional Chaos

In pure mathematics, nothing can stop us from dreaming up hyperspace universes containing as many dimensions as we desire. Imagine, for example, *Cartesian 25-space* in which coordinates take the form of *ordered 25-tuples* $(x_1,x_2,x_3,...,x_{25})$, none of which represent time. Alternatively, we might allow Cartesian 25-space to contain 24 spatial dimensions and one time dimension. Then we would define the coordinates of a point as an ordered 25-tuple of the form $(x_1,x_2,x_3,...,x_{24},t)$.

Some *cosmologists*—scientists who explore the origin, structure, and evolution of the cosmos—have suggested that our universe contained many more than three spatial dimensions in its first few moments of existence, billions of years ago. According to this hypothesis, we cannot represent all of these dimensions using Cartesian coordinates. Some of the axes are "curled up" as if wrapped around tiny bubbles.

A few intrepid mathematicians play with objects that seem to occupy two dimensions when imagined in a certain way, yet occupy three dimensions when imagined in a different way. Some inquisitive people ask questions such as, "How many dimensions exist in the complicated surface of a theoretical *foam*, assuming that each individual bubble constitutes a sphere of arbitrarily tiny size and with an infinitely thin 2D surface? Two dimensions? Three? How about two and a half dimensions?"

TIP *As you can doubtlessly imagine by now, dimensional scenarios can get a lot more complicated than anything we've dealt with here. Think about the possible ways in which a 4D parallelepiped might exist, or a 4D sphere. How about a 5D sphere or a 7D ellipsoid? Let your mind roam free.*

Distance Formulas

In n-dimensional Cartesian space, we can calculate the shortest distance between any two known points using a formula similar to the distance formulas for Cartesian two-space and three-space. The outcome of our arithmetic represents the length of a straight line segment connecting the two points. Consider two points P and Q in Cartesian n-space whose coordinates are

$$P = (x_1, x_2, x_3, \dots, x_n)$$

and

$$Q = (y_1, y_2, y_3, \dots, y_n)$$

We can find the length of the shortest possible path between P and Q, written $|PQ|$, with the formula

$$|PQ| = [(y_1 - x_1)^2 + (y_2 - x_2)^2 + (y_3 - x_3)^2 + \cdots + (y_n - x_n)^2]^{1/2}$$

or the alternative

$$|PQ| = [(x_1 - y_1)^2 + (x_2 - y_2)^2 + (x_3 - y_3)^2 + \cdots + (x_n - y_n)^2]^{1/2}$$

PROBLEM 11-2

Find the distance $|PQ|$ between the points $P = (4, -6, -3, 0)$ and $Q = (-3, 5, 0, 8)$ in Cartesian four-space. Assume the coordinate values to be exact. Round off the answer to two decimal places.

SOLUTION

Let's assign the numbers in the ordered quadruples the following values according to the formatting of the above formulas. For P, we have

$$x_1 = 4$$
$$x_2 = -6$$
$$x_3 = -3$$
$$x_4 = 0$$

For Q, we have

$$y_1 = -3$$
$$y_2 = 5$$
$$y_3 = 0$$
$$y_4 = 8$$

Now, we can plug these values into either of the above two distance formulas. If we use the first one, we obtain

$$|PQ| = \{(-3-4)^2 + [5-(-6)]^2 + [0-(-3)]^2 + (8-0)^2\}^{1/2}$$
$$= [(-7)^2 + 11^2 + 3^2 + 8^2]^{1/2}$$
$$= (49 + 121 + 9 + 64)^{1/2}$$
$$= 243^{1/2}$$
$$= 15.59$$

PROBLEM 11-3

How many vertices does a tesseract have?

SOLUTION

Imagine a tesseract as a 3D cube that lasts for a length of time equivalent to the linear span of each edge. When we think of a tesseract this way, and if we think of time as "flowing upward" from the past toward the future, the tesseract has a "bottom" that represents the instant that it appears and a "top" that represents the instant that it vanishes. The "bottom" and the "top" of the tesseract, thereby defined, form two separate cubes. We know that a cube has eight vertices. In the tesseract, we observe twice as many

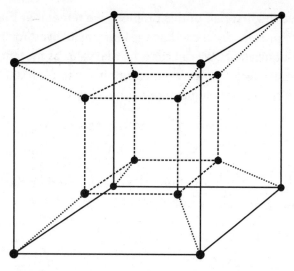

FIGURE 11-7 · The cube-within-a-cube portrayal of a tesseract clarifies the fact that the figure has 16 vertices.

vertices as we do in a cube, because we join two cubes with line segments between corresponding pairs of vertices. The eight vertices of the "bottom" cube and the eight vertices of the "top" cube connect pairwise with line segments that run through time.

TIP   *We can think of a "dimensionally reduced" tesseract as a cube-within-a-cube as shown in Fig. 11-7. Illustrators sometimes use this trick in a 3D attempt to portray a 4D tesseract. We don't get a true picture this way, of course, because the "inner" and the "outer" cubes in a real tesseract are the same size. But this rendition demonstrates the fact that a tesseract has 16 vertices. We can simply count them!*

PROBLEM 11-4

What's the 4D hypervolume, V_{4D}, of a rectangular four-prism consisting of a 3D cube measuring exactly 1 meter on each edge, that "lives" for exactly 1 second, and that does not move? Express the answer in quartic kilometer-equivalents and in quartic microsecond-equivalents.

SOLUTION

We must find the 4D hypervolume of a 3D cube measuring $1 \times 1 \times 1$ meter (whose 3D volume therefore equals 1 cubic meter) that exists for precisely 1 second.

To solve the first half of this problem, we remember that light travels 300,000 kilometers per second, so the four-prism "lives" for 300,000, or 10^5, kilometer-equivalents. We can consider that value as the length of the four-prism. Its cross section is a cube measuring 1 meter, or 0.001 kilometer, on each edge, so the 3D volume of this cube equals

$$0.001 \times 0.001 \times 0.001 = 0.000000001$$
$$= 10^{-9} \text{ cubic kilometer}$$

Therefore, the 4D hypervolume (V_{4D}) of the rectangular four-prism in quartic kilometer-equivalents is

$$V_{4D} = 300,000 \times 0.000000001$$
$$= 3 \times 10^5 \times 10^{-9}$$
$$= 3 \times 10^{-4}$$
$$= 0.0003 \text{ quartic kilometer-equivalent}$$

To solve the second half of the problem, let's note that in 1 microsecond (0.000001 second), a ray of light travels 300 meters, so it takes light 1/300 of a microsecond to travel 1 meter. The 3D volume of the cube is therefore

$$(1/300)^3 = 1/27,000,000$$
$$= 0.00000003704$$
$$= 3.704 \times 10^{-8} \text{ cubic microsecond-equivalent}$$

The cube exists for 1 second, which equals 1,000,000, or 10^6, microseconds. Therefore, the 4D hypervolume V_{4D} of the rectangular four-prism in quartic microsecond-equivalents is

$$V_{4D} = 0.00000003704 \times 1,000,000$$
$$= 3.704 \times 10^{-8} \times 10^6$$
$$= 3.704 \times 10^{-2}$$
$$= 0.03704 \text{ quartic microsecond-equivalent}$$

PROBLEM 11-5

Suppose that the four-prism described in Problem 11-4 moves, during its brief existence, at a speed of 270,000 kilometers per second relative to an observer. What is its 4D hypervolume (V_{4D}) as seen by that observer? Express the answer in quartic kilometer-equivalents and in quartic microsecond-equivalents.

SOLUTION

The object moves at 270,000/300,000, or 9/10, of the speed of light relative to the observer. If we let s represent its speed, then $s/c = 0.9$ and $s^2/c^2 = 0.81$. We must multiply the answers to the previous problem by the factor

$$(1 - s^2/c^2)^{1/2} = (1 - 0.81)^{1/2}$$
$$= 0.19^{1/2}$$
$$= 0.436$$

When we apply this conversion factor to the solutions we got for Problem 11-4, we obtain the 4D hypervolume values

$$V_{4D} = 0.0003 \times 0.436$$
$$= 0.000131 \text{ quartic kilometer-equivalent}$$

and

$$V_{4D} = 0.03704 \times 0.436$$
$$= 0.0161 \text{ quartic microsecond-equivalent}$$

Parallel Principle Revisited

All of the theorems in conventional geometry derive from five *axioms*, also called *postulates*, originally formalized by the Greek mathematician *Euclid of Alexandria* who lived in the third century B.C. Everything that we've done in this book so far—even the theoretical problems involving four dimensions—has evolved and worked out according to Euclid's five axioms. We've dealt exclusively with so-called *Euclidean geometry*. However, other "flavors" of geometry exist, in which Euclid's axioms do not necessarily all hold true. Mathematicians call any such discipline *non-Euclidean geometry*.

Euclid's Axioms

Let's examine the statements that Euclid regarded as self-evident truths. We'll modify Euclid's original wording slightly, so as to make the passages sound sensible in today's language. Figure 11-8 shows examples of each postulate.

- We can connect any two points P and Q with a straight line segment (Fig. 11-8A).
- We can extend any straight line segment indefinitely and continuously to form a straight line (Fig. 11-8B).
- Given any point P, we can define a circle having a specific radius r with P at its center (Fig. 11-8C).

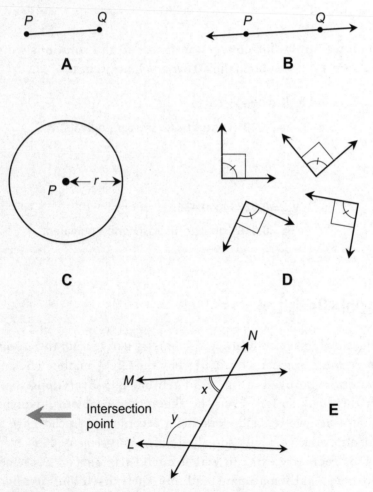

FIGURE 11-8 · Euclid's original five axioms. See text for discussion.

- All right angles are congruent to one another; that is, all right angles have equal measures (Fig. 11-8D).
- Consider two lines L and M that lie in the same plane, and a transversal line N that crosses them both. Suppose that the measure of the acute or right angle between M and N (x as shown in Fig. 11-8E) and the measure of the obtuse or right angle between L and N (y as shown in Fig. 11-8E) add up to something less than 180° (π rad). In that case, lines L and M intersect at some point on the same side of line N as the adjacent angles x and y lie.

The Fifth Postulate

The last axiom stated above has become known as *Euclid's fifth postulate*. It's logically equivalent to the following statement called the *parallel postulate*:

- Let L represent a straight line. Let P represent a point that does not lie on L. There exists one and only one straight line M, in the plane defined by line L and point P, that passes through P and runs parallel to L.

This axiom—and in particular its truth or untruth—has received enormous attention from geometers over the last few hundred years. If we deny the parallel postulate, we end up with a system of geometry that "works" just as well as traditional plane geometry does. Some people find such "geometries" strange, but they're logically sound in the sense that contradictions don't arise. We can deny the truth of the parallel postulate in either of two ways:

- There exists no line M through point P that runs parallel to line L.
- There exist two or more lines M_1, M_2, M_3, ... through point P that run parallel to line L.

When we replace Euclid's original parallel postulate with either of the foregoing two variants, we get a system of non-Euclidean geometry. In the 2D case, we find ourselves confined to a *non-Euclidean surface*. Visually, such a surface looks warped or curved.

Geodesics

In a non-Euclidean universe, we must modify the concept of "straightness" and, in particular, the notion of what constitutes a "line." Instead of thinking about "straight lines" or "straight line segments," we must think about *geodesics*.

Imagine two distinct points P and Q on a non-Euclidean surface. The *geodesic segment* or *geodesic arc* connecting P and Q is the set of points representing the shortest possible path between P and Q that lies entirely on the surface. If we extend a geodesic arc indefinitely in either direction on the surface beyond P and Q, we obtain the complete *geodesic* within which the arc lies.

Still Struggling

Do you have trouble imagining a geodesic arc in your "mind's eye"? Think about the path that a thin ray of light would follow between two points if confined to a certain 2D universe. The extended geodesic conforms to the path that the ray would take if allowed to travel over the surface forever without striking any obstructions. On the surface of the earth, a geodesic arc is the path that an airline pilot takes when flying from one place to another far away, such as from Moscow, Russia to Tokyo, Japan (neglecting takeoff and landing patterns and assuming that the pilot doesn't have to adjust the course to avoid storms or hostile air space).

Modified Parallel Postulate

When we restate the parallel postulate as it applies to both Euclidean and non-Euclidean surfaces, we must replace the term "line" with "geodesic." When two geodesics G and H lie on the same surface X but fail to intersect at any point on X, we say that G and H constitute a pair of *parallel geodesics* on X. Let G represent a geodesic, let X represent a surface, and let P represent a point that does not lie on G. Then one of the following three situations holds true:

- There exists exactly one geodesic H on X that passes through P and runs parallel to G.
- There exists no geodesic H on X that passes through P and runs parallel to G.
- There exist two or more geodesics H_1, H_2, H_3, ... on X that pass through P and run parallel to G.

No Parallel Geodesics

Now imagine a universe U in which no two geodesics ever run parallel to each other. In the universe U, if we extend two geodesic arcs that "look" parallel on a local scale far enough in both directions off their ends, they'll eventually intersect at some point in U. In a universe of this sort, we must employ a system of *elliptic geometry*, also known as *Riemannian geometry* (named after *Bernhard Riemann*, a nineteenth-century German mathematician). When there exist no pairs of parallel geodesics in a particular universe U, we say that U has *positive curvature*. Two-space universes with positive curvature include the surfaces of spheres, *oblate* (flattened) *spheres*, and *ellipsoids*.

Figure 11-9 illustrates a sphere with a triangle and a quadrilateral on its surface. The sides of polygons in non-Euclidean geometry always constitute geodesic arcs, just as, in Euclidean geometry, they always constitute straight line segments. The interior angles of the triangle and the quadrilateral in Fig. 11-9 add up to more than 180° (π rad) and more than 360° (2π rad), respectively. The measures of the interior angles of an n-sided polygon on a Riemannian surface always sum up to something more than the sum of the measures of the interior angles of an n-sided polygon on a Euclidean (flat) plane.

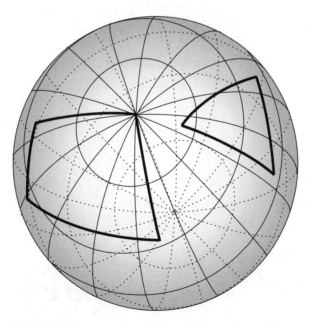

FIGURE 11-9 · A surface with positive curvature, in this case a sphere, showing a triangle and a quadrilateral whose sides constitute geodesics.

TIP *On the surface of the earth, all the lines of longitude, called* meridians, *are geodesics. So is the equator. But latitude circles other than the equator, called* parallels, *are not geodesics. For example, the equator and the parallel representing 10° north latitude don't intersect at any point, but they aren't both geodesics.*

More Than One Parallel Geodesic

Consider a surface on which we can have two or more geodesics that pass through a point and run parallel to a given geodesic. This form of non-Euclidean geometry is known as *hyperbolic geometry*. Some mathematicians call it *Lobachevskian geometry* (named after *Nikolai Lobachevsky*, a nineteenth-century Russian mathematician). A Lobachevskian universe exhibits so-called *negative curvature*. Two-space universes with negative curvature include extended saddle-shaped and funnel-shaped surfaces.

Figure 11-10 shows a negatively curved surface containing a triangle and a quadrilateral. On this surface, the interior angles of the triangle and the

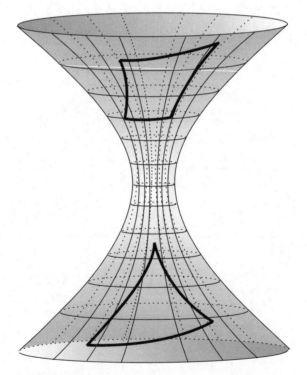

FIGURE 11-10 · An example of a surface with negative curvature, showing a triangle and a quadrilateral whose sides constitute geodesics.

quadrilateral add up to less than 180° (π rad) and 360° (2π rad), respectively. The measures of the interior angles of a polygon on a Lobachevskian surface always sum up to something less than the sum of the measures of the interior angles of a similar polygon on a Euclidean plane.

Warped Space

The observable universe seems Euclidean to all casual observers. If we use lasers to "construct" polygons and then measure their interior angles with precision lab equipment, we'll always find that the angle measures add up according to the rules of Euclidean geometry. The conventional formulas for the volumes of solids such as the pyramid, cube, and sphere hold perfectly, as far as we can tell. Now imagine a 3D space in which these rules fail! If we could find such a continuum, we would call it *curved 3D space*, *warped 3D space*, or *non-Euclidean 3D space*.

Gravity Warps Space

In the 1900s, shortly after Einstein published the details of his *general theory of relativity*, astronomers and cosmologists began to look for evidence that the three-space in which we live is not perfectly Euclidean. Their efforts reaped fascinating results. Gravitational fields produce effects on light beams that suggest Lobachevskian warping—negative curvature—of three-space. Under ordinary circumstances, the departure from Euclidean perfection is too small to notice, so we never suspect it. However, astronomers have observed the effects of such curvature using sensitive equipment when looking at certain celestial objects.

Astronomers conducted several experiments in the years following the publication of Einstein's general theory, scrutinizing the behavior of light rays from distant stars as the rays passed close to the sun during solar eclipses. The goal: Find out whether or not the sun's gravitational field, which attains considerable intensity near the sun's surface, bends light rays in the way that we should expect if space has negative curvature near the sun. Early in the twentieth century, Albert Einstein predicted that such bending could be observed and measured. He calculated the expected angular changes that astronomers would see in the positions of distant stars as the sun passes almost directly in front of them. Repeated observations verified Einstein's predictions, not only as to the existence of the spatial curvature, but also as to its extent. As the distance from the sun increases, the spatial warping decreases. The greatest amount of light-beam bending occurs when the rays from a distant star graze the sun's surface.

In another experiment, astronomers have observed the light rays from a distant, brilliant object called a *quasar* as a compact, dark, intense source of gravitation (known as a *black hole*) passes between the quasar and our solar system. The light-bending is much greater near this type of object than is the case near the sun. The apparent black hole bends the rays from the distant quasar to the extent that multiple images of the quasar appear (with the black hole presumably at the center). One peculiar example, in which four images of the quasar appear, has been called a *gravitational light cross*.

The "Hyperfunnel"

We can compare curvature of space in the presence of a strong gravitational field to the shape of a funnel (Fig. 11-11), except that the surface of the funnel has three dimensions rather than two, and the entire object exists in four-space

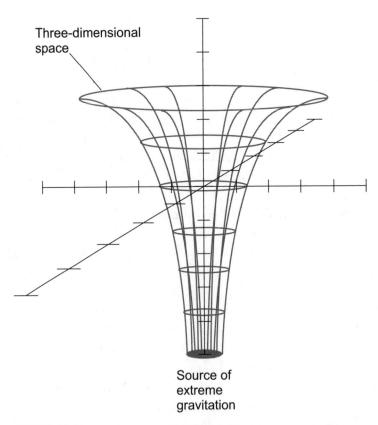

FIGURE 11-11 · An intense source of gravitation produces negative curvature, or warping, of space in its immediate vicinity.

rather than three-space. When we define the fourth dimension as time, we find that time "flows" more slowly in a gravitational field than it does in interplanetary space, far removed from significant sources of gravitation. This effect, like the Lobachevskian curvature of space, has been experimentally observed.

The shortest path in physical three-space between any two points near a gravitational source lies along a geodesic, not along a straight line. Curvature of space caused by gravitational fields increases the distances between points in the vicinity of the source of the gravitation, compared with the situation if the gravitational source were not there. The shortest path between any two points in non-Euclidean space invariably exceeds the path length that we would observe if the space between the points were Euclidean.

As the intensity of the gravitation increases, the extent of the spatial curvature also increases. However, some effect theoretically occurs no matter how weak the gravitation. Some cosmic warping occurs in the space around the earth, in the space around your body, and even in the space around each atom in your body.

Still Struggling

Does the entire universe, containing all the stars, galaxies, quasars, and other stuff that exists, possess a geometric shape that results from the combined gravitational effect of all matter? If so, is the general contour of space Riemannian, Lobachevskian, or Euclidean? I don't think anybody knows for sure. Do you?

QUIZ

Refer to the text in this chapter if necessary. A good score is eight correct. Answers are in the back of the book.

1. Using the speed of light in free space as the basis for conversion, what's the distance equivalent of 1 minute?
   A. 5000 kilometers
   B. 500,000 kilometers
   C. 1,800,000 kilometers
   D. 18,000,000 kilometers

2. Using the speed of light in free space as the basis for conversion, what's the time equivalent of 150 meters?
   A. 500 nanoseconds (0.0000005 second)
   B. 200 nanoseconds (0.0000002 second)
   C. 500 microseconds (0.0005 second)
   D. 200 microseconds (0.0002 second)

3. Figure 11-12 illustrates a light cone in dimensionally reduced time-space, along with four hypothetical paths P, Q, R, and S for objects traveling in

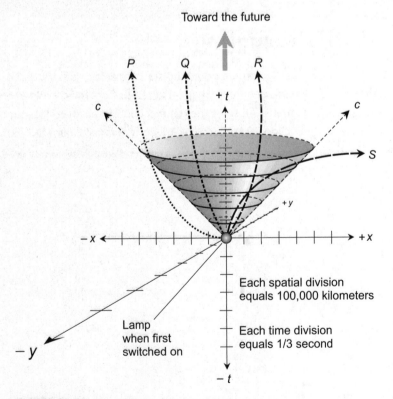

FIGURE 11-12 · Illustration for Quiz Question 3.

that space. Which, if any, of these paths could a physical object actually follow?

A. *P* and *Q*
B. *Q* and *R*
C. *P* and *S*
D. None of the above

4. On the surface of a sphere, the measure of each interior angle of a regular hexagon would

A. be less than 120°.
B. equal 120°.
C. exceed 120°.
D. be impossible to define.

5. Figure 11-13 shows a sphere, along with a specific circle called *C* (heavy solid curve) on its surface. The radius of *C* equals 200 units as measured over the surface of the sphere along a geodesic arc. Based on this information, we know that the circumference of *C* must

A. exceed 400π units.
B. be less than 400π units.
C. equal 400π units.
D. be impossible to define.

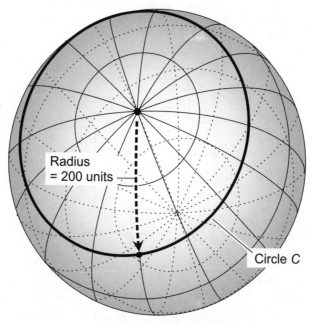

FIGURE 11-13 · Illustration for Quiz Question 5.

6. What's the distance between the origin and the point (1,1,1,1) in Cartesian four-space? Assume the coordinate values to be exact.

A. The cube root of 8 units
B. The square root of 2 units
C. 2 units
D. The square root of 8 units

7. Imagine a rectangular prism that measures exactly 60 meters high, 120 meters wide, and 300 meters deep. Suppose that it forms from nothing, exists in free space for exactly 0.01 second, and then vanishes. What's its hypervolume in quartic kilometer-equivalents? Assume that the free-space speed of light equals exactly 3×10^5 kilometers per second.

A. 6.48 quartic kilometer-equivalents
B. 2.40 quartic kilometer-equivalents
C. 1.80 quartic kilometer-equivalents
D. 2.16 quartic kilometer-equivalents

8. Imagine a rectangular prism that measures exactly 60 meters high, 120 meters wide, and 300 meters deep. Suppose that it forms from nothing, exists in free space for exactly 0.01 second, and then vanishes. What's its hypervolume in quartic microsecond-equivalents? Assume that the free-space speed of light equals exactly 3×10^5 kilometers per second.

A. 200 quartic microsecond-equivalents
B. 333 quartic microsecond-equivalents
C. 667 quartic microsecond-equivalents
D. 800 quartic microsecond-equivalents

9. Figure 11-14 portrays a tesseract in "dimensionally reduced" form. As we've already learned, this 4D figure has 16 vertices. How many line-segment edges does the figure have?

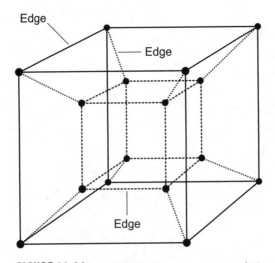

FIGURE 11-14 · Illustration for Quiz Questions 9 and 10.

A. 16
B. 24
C. 32
D. 40

10. Imagine that the tesseract of Fig. 11-14 has a hypervolume of exactly 4096 quartic units. How long is each edge? Remember that in a true tesseract, all the edges have equal length (despite the distorted appearance of this illustration).

A. 16 units
B. The square root of 128 units
C. The cube root of 2048 units
D. 8 units

Test: Part II

Do not refer to the text when taking this test. You may draw diagrams or use a calculator if necessary. A good score is at least 38 correct. Answers are in the back of the book. It's best to have a friend check your score the first time, so you won't memorize the answers if you want to take the test again.

1. Imagine that two planes intersect in a straight line. We can express the angle at which the planes intersect in two ways: as an acute angle *u* or as an obtuse angle *v*. If we measure both angles in degrees, then

   A. $u = 90° - v$.
   B. $u = 120° - v$.
   C. $u = 180° - v$.
   D. $u = 270° - v$.
   E. $u = 300° - v$.

2. The equation of the line $x = 0$, as expressed in the Cartesian plane, translates to the polar-coordinate equation

   A. $\theta = \pi/8$.
   B. $\theta = \pi/4$.
   C. $\theta = \pi/2$.
   D. $\theta = 2\pi/3$.
   E. $\theta = 3\pi/4$.

3. The equation of the line $y = -x$, as expressed in the Cartesian plane, translates to the polar-coordinate equation

   A. $\theta = \pi/8$.
   B. $\theta = \pi/4$.
   C. $\theta = \pi/2$.
   D. $\theta = 2\pi/3$.
   E. $\theta = 3\pi/4$.

4. What's the sum of the Cartesian three-space vectors u = (2,3,4) and v = (−4,−3,−2)?

   A. u + v = (−2,0,2)
   B. u + v = (2,0,−2)
   C. u + v = (6,6,6)
   D. u + v = (−6,−6,−6)
   E. u + v = (0,0,0)

5. Figure Test II-1 illustrates various geometric shapes in a Euclidean plane. The only difference among the versions shown in Figs. Test II-1A, B, and C involves the extent to which the figures include their boundaries. Based on this information, what can we say about the three different scenarios shown in Figs. Test II-1A, B, and C?

   A. The figures in Fig. Test II-1A have greater perimeters than their counterparts in Fig. Test II-1B, which in turn have greater perimeters than their counterparts in Fig. Test II-1C. However, the figures in Fig. Test II-1A have the same interior areas as their counterparts in Fig. Test II-1B, which in turn have the same interior areas as their counterparts in Fig. Test II-1C.
   B. The figures in Fig. Test II-1A have greater perimeters than their counterparts in Fig. Test II-1B, which in turn have greater perimeters than their counterparts in Fig. Test II-1C. Also, the figures in Fig. Test II-1A have greater interior areas than

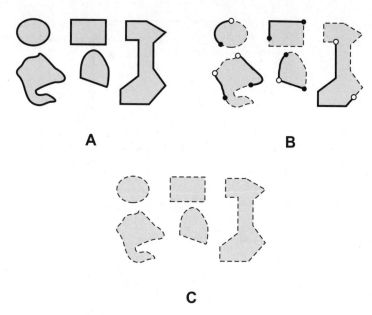

A   B

C

FIGURE TEST II-1 · Illustration for Part II Test Question 5.

their counterparts in Fig. Test II-1B, which in turn have greater interior areas than their counterparts in Fig. Test II-1C.

C. The figures in Fig. Test II-1A have greater interior areas than their counterparts in Fig. Test II-1B, which in turn have greater interior areas than their counterparts in Fig. Test II-1C. However, the figures in Fig. Test II-1A have the same perimeters as their counterparts in Fig. Test II-1B, which in turn have the same perimeters as their counterparts in Fig. Test II-1C.

D. The figures in Fig. Test II-1A have the same interior areas as their counterparts in Fig. Test II-1B, which in turn have the same interior areas as their counterparts in Fig. Test II-1C. Also, the figures in Fig. Test II-1A have the same perimeters as their counterparts in Fig. Test II-1B, which in turn have the same perimeters as their counterparts in Fig. Test II-1C.

E. We cannot make any of the above general statements.

6. **In three-space, whenever two flat planes intersect but do not actually coincide, their intersection can take the form of a**
    A. point or a straight line.
    B. point or a straight ray.
    C. straight ray or a straight line.
    D. point, a straight ray, or a straight line.
    E. straight line only.

7. **Figure Test II-2 illustrates three planes *X*, *Y*, and *Z*. Planes *X* and *Y* intersect in a straight line *L*. Planes *X* and *Z* intersect in a straight line *M*. Lines *L* and *M* run parallel to each other. Line *PQ* lies in plane *X* and runs perpendicular to both**

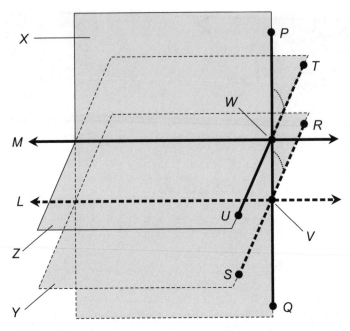

FIGURE TEST II-2 · Illustration for Part II Test Question 7.

lines *L* and *M*. Line *RS* lies in plane *Y* and runs perpendicular to both lines *L* and *PQ*. Line *TU* lies in plane *Z* and runs parallel to line *RS*. Angle *TWP* has the same measure as angle *RVP* (both angles are denoted by dashed arcs). Based on all this information, what can we say about planes *Y* and *Z*?

A. They're parallel to each other.
B. They're skew to each other.
C. They're normal to each other.
D. They must intersect at some point not shown here.
E. They must intersect at some line not shown here.

8. What's the surface area of a rectangular prism that measures 4 inches high, 5 inches wide, and 7 inches deep?

A. 32 square inches
B. The square root of 140 square inches
C. The square root of 166 square inches
D. 140 square inches
E. 166 square inches

9. What's the volume of a rectangular prism that measures 4 inches high, 5 inches wide, and 7 inches deep?

A. 32 cubic inches
B. The square root of 140 cubic inches
C. The square root of 166 cubic inches
D. 140 cubic inches
E. 166 cubic inches

10. In celestial coordinates, *declination* is the equivalent of
    A. right ascension.
    B. celestial latitude.
    C. celestial longitude.
    D. elevation.
    E. azimuth.

11. In the context of terrestrial (earth-based) navigation, the term *declination* can refer to something entirely different than its meaning in the context of celestial coordinates: the angular difference between
    A. magnetic north and geographic north.
    B. azimuth and elevation.
    C. right ascension and celestial longitude.
    D. the vernal equinox and the zenith.
    E. the zenith and the celestial latitude.

12. Consider a flat plane in three-space, and a straight line that does not intersect the plane at any point. In this situation, the line and the plane are
    A. orthogonal.
    B. perpendicular.
    C. non-Euclidean.
    D. normal.
    E. parallel.

13. Suppose that in the Cartesian coordinate plane, a certain vector u begins (originates) at the point (−5,8) and ends (terminates) at the point (3,1). Which of the following ordered pairs represents u in standard form?
    A. $\mathbf{u} = (-8,7)$
    B. $\mathbf{u} = (8,-7)$
    C. $\mathbf{u} = (-2,9)$
    D. $\mathbf{u} = (8,9)$
    E. $\mathbf{u} = (-15,8)$

14. Suppose that in the Cartesian plane, a certain vector v begins at the point (3,1) and ends at the point (−5,8), exactly the opposite state of affairs from the situation described in Question 13. What's v in standard form?
    A. $\mathbf{v} = (-8,7)$
    B. $\mathbf{v} = (8,-7)$
    C. $\mathbf{v} = (-2,9)$
    D. $\mathbf{v} = (8,9)$
    E. $\mathbf{v} = (-15,8)$

15. What's the four-dimensional (4D) hypervolume, in *quartic units*, of a tesseract that measures exactly 5 units on each edge?
    A. 1024 quartic units
    B. 625 quartic units
    C. 125 quartic units
    D. 25 quartic units
    E. 20 quartic units

16. What's the 4D hypervolume, in *quartic meter-equivalents*, of a rectangular four-prism consisting of a three-dimensional (3D) cube measuring exactly 2 meters on each edge, that "lives" for exactly 2 seconds, and that does not move? Assume that the speed of light in free space equals 3.00×10^8 meters per second.

  A. 4.8×10^9 quartic meter-equivalents
  B. 2.4×10^9 quartic meter-equivalents
  C. 1.2×10^9 quartic meter-equivalents
  D. 9.6×10^8 quartic meter-equivalents
  E. 4.8×10^8 quartic meter-equivalents

17. The point $(-1,-1)$ in the Cartesian xy-plane corresponds to one of the following points in mathematician's (θ, r) polar coordinates (MPC). Which point? Remember that the 1/2 power of a number equals the positive square root of that number.

  A. $(\pi/4, 2^{1/2})$
  B. $(\pi/2, 2^{1/2})$
  C. $(3\pi/4, 2^{1/2})$
  D. $(5\pi/4, 2^{1/2})$
  E. $(7\pi/4, 2^{1/2})$

18. Figure Test II-3 illustrates a slant circular cylinder. What's the volume of the enclosed solid? Assume that the base radius, the top radius, and the height have exactly the values shown. Here's a hint: The area enclosed by a circle equals π times the square of its radius. Here's another hint: To find the area of a cylinder (whether it's slanted or not), multiply its height by the enclosed area of its base.

  A. 84 cubic units
  B. 168 cubic units
  C. 336 cubic units
  D. 475 cubic units
  E. 672 cubic units

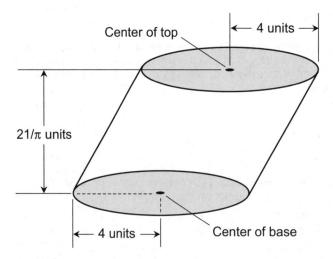

FIGURE TEST II-3 · Illustration for Part II Test Question 18.

19. Considered with respect to the speed of light in free space (3.00×10^8 meters per second), the distance from the earth to the sun (1.50×10^8 kilometers) represents a time differential of

   A. 33 minutes and 20 seconds
   B. 16 minutes and 40 seconds.
   C. 8 minutes and 20 seconds.
   D. 4 minutes and 10 seconds.
   E. 2 minutes and 5 seconds.

20. If we double the radius of a slant circular cone's base but do not change the cone's height, the volume of the enclosed solid increases by a factor of

   A. the fourth root of 2.
   B. the cube root of 2.
   C. the square root of 2.
   D. 2.
   E. 4.

21. If we double the height of a slant circular cone but do not change the cone's base radius, the volume of the enclosed increases by a factor of

   A. the fourth root of 2.
   B. the cube root of 2.
   C. the square root of 2.
   D. 2.
   E. 4.

22. What's the distance between (0,0,0,0,0,0,0,0) and (−1,−1,−1,−1,−1,−1,−1,−1) in Cartesian eight-space?

   A. 1 unit
   B. The eighth root of 2 units
   C. The eighth root of −1 unit
   D. The square root of 8 units
   E. We can't define it.

23. If we increase the volume of a perfect cube by a factor of 4, its surface area increases by a factor of the

   A. cube root of 16.
   B. cube root of 32.
   C. cube root of 64.
   D. square root of 8.
   E. square root of 32.

24. What's the six-dimensional (6D) hypervolume, in *hexic meters*, of a six-cube measuring 1 meter on each edge?

   A. 1 hexic meter
   B. 6 hexic meters
   C. 36 hexic meters
   D. 216 hexic meters
   E. 1296 hexic meters

25. If we triple the radius of a sphere, its surface area increases by a factor of
    A. 27.
    B. 9.
    C. 3.
    D. the square root of 3.
    E. the cube root of 3.

26. If we triple the radius of a sphere, its enclosed volume increases by a factor of
    A. 27.
    B. 9.
    C. 3.
    D. the square root of 3.
    E. the cube root of 3.

27. Figure Test II-4 shows two vectors g and h in Cartesian three-space. Which of the following statements holds true for them?
    A. $\mathbf{g} \bullet \mathbf{h} = (0,-10,25)$
    B. $\mathbf{g} \bullet \mathbf{h} = (-1,3,10)$
    C. $\mathbf{g} \bullet \mathbf{h} = 0$ (the scalar 0)
    D. $\mathbf{g} \bullet \mathbf{h} = 15$ (the scalar 15)
    E. $\mathbf{g} \bullet \mathbf{h} = \mathbf{0}$ (the zero vector)

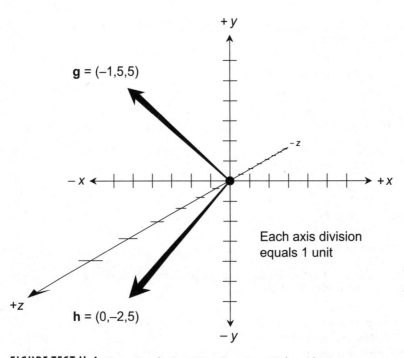

FIGURE TEST II-4 · Illustration for Part II Test Questions 27 through 29.

28. Which of the following statements holds true for the vectors of Fig. Test II-4?

  A. $g+h = (0,-10,25)$
  B. $g+h = (-1,3,10)$
  C. $g+h = 0$ (the scalar zero)
  D. $g+h = 15$
  E. $g+h = 0$ (the zero vector)

29. We can represent any straight line in Cartesian *xyz*-space as a symmetric equation of the form

$$(x - x_0)/a = (y - y_0)/b = (z - z_0)/c$$

  where *x*, *y*, and *z* represent the variables; the ordered triple (x_0, y_0, z_0) tells us the coordinates of a specific point on the line; and *a*, *b*, and *c* represent the line's direction numbers. Consider a line *L* connecting the two points at the nonorigin (terminating) ends of the vectors g and h shown in Fig. Test II-4. What are the direction numbers of *L* in the form of an ordered triple (a,b,c)? Here's a hint: Determine the standard form of a vector that originates at the nonorigin end of h and terminates at the nonorigin end of g.

  A. $(a,b,c) = (-1,7,0)$
  B. $(a,b,c) = (0,-10,25)$
  C. $(a,b,c) = (-1,3,10)$
  D. $(a,b,c) = (0,0,0)$
  E. We can't define them.

30. The faces (including the base) of a tetrahedron are all

  A. triangles.
  B. squares.
  C. rectangles.
  D. rhombuses.
  E. parallelograms.

31. Two distinct, flat half planes in three-space run parallel to each other if and only if the complete planes in which they lie intersect

  A. nowhere.
  B. in a single point.
  C. in a straight ray.
  D. in a straight line.
  E. in either a straight ray or a straight line.

32. In Fig. Test II-5, suppose that we let each radial division (distance outward from one of the concentric circles to the next one) represent exactly 1 unit. In that case, what's the equation of the spiral?

  A. $r = 2\theta$
  B. $r = 2\theta/3$
  C. $r = 2\theta/\pi$
  D. $r = \theta/3$
  E. $r = \theta/\pi$

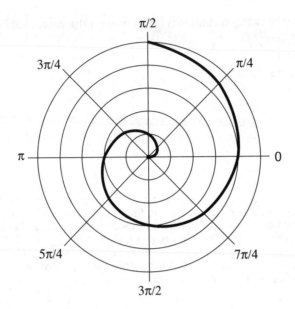

FIGURE TEST II-5 · Illustration for Part II Test Questions 32 and 33.

33. In Fig. Test II-5, suppose that we let each radial division represent exactly π units. In that case, what's the equation of the spiral?

 A. $r = 2\theta$
 B. $r = 2\theta/3$
 C. $r = 2\theta/\pi$
 D. $r = \theta/3$
 E. $r = \theta/\pi$

34. Figure Test II-6 illustrates a 3D set of

 A. Cartesian coordinates.
 B. cylindrical coordinates.
 C. terrestrial coordinates.
 D. spherical coordinates.
 E. elliptical coordinates.

35. What's the dot product of the Cartesian-plane vectors **q** = (0,5) and **r** = (−5,0)?

 A. The zero vector
 B. The vector (−5,5)
 C. The scalar quantity 25
 D. The scalar quantity 0
 E. A vector that lies outside the plane containing **q** and **r**

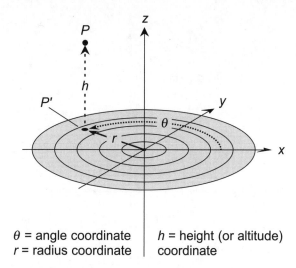

θ = angle coordinate h = height (or altitude)
r = radius coordinate   coordinate

FIGURE TEST II-6 · Illustration for Part II Test Question 34.

36. **What's the cross product of the Cartesian-plane vectors q = (0,5) and r = (−5,0)?**
   A. The zero vector
   B. The vector (−5,5)
   C. The scalar quantity 25
   D. The scalar quantity 0
   E. A vector that lies outside the plane containing **q** and **r**

37. **Imagine that two planes intersect at an angle of 60°, representing the more common of two ways in which we can express the intersection angle. What's a less common, but still technically valid, expression for the intersection angle in the same situation?**
   A. −70°
   B. −30°
   C. 100°
   D. 120°
   E. 150°

38. **If we double the length of one semiaxis in an ellipsoid while not changing the lengths of the other two semiaxes, we increase the volume of the enclosed solid by a factor of**
   A. the square root of 2.
   B. 2.
   C. the square root of 8.
   D. 4.
   E. 8.

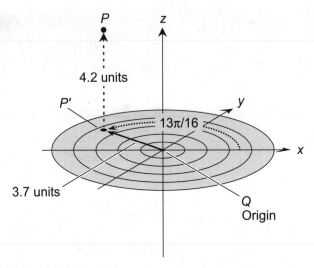

FIGURE TEST II-7 · Illustration for Part II Test Question 39.

39. Imagine that in the scenario of Fig. Test II-7, we construct line segment *PQ*. What's its length to the nearest tenth of a unit?
   A. 5.6 units
   B. 6.2 units
   C. 7.0 units
   D. 7.9 units
   E. We need more information to answer this question.

40. By convention, we can express the distance between two flat, parallel planes along any line that
   A. runs parallel to both planes.
   B. runs parallel to only one of the planes.
   C. runs normal to both planes.
   D. intersects both planes.
   E. intersects only one of the planes.

41. At a minimum, how many distinct points do we need to uniquely define a flat plane in three-space?
   A. None
   B. One
   C. Two
   D. Three
   E. Four

42. Figure Test II-8 is a time-space graph of the earth (small black dot) as it revolves around the sun. What's the time-space value of d_t in second-equivalents, assuming

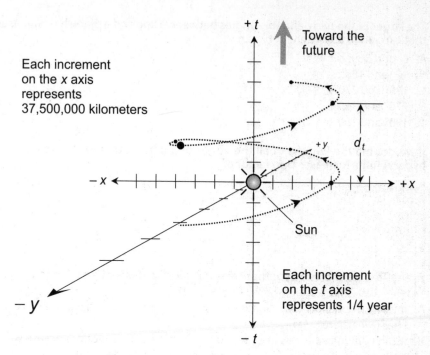

FIGURE TEST II-8 · Illustration for Part II Test Questions 42 and 43.

that we consider the earth's orbit as a perfect circle and the sun as an absolutely stationary point of reference? Assume that the speed of light in free space is 3.00×10^8 meters per second. Also assume that 1 year equals exactly 365 days, each of which contains exactly 24 hours.

A. 6.31×10^7 second-equivalents

B. 3.15×10^7 second-equivalents

C. 8.64×10^6 second-equivalents

D. 4.32×10^6 second-equivalents

E. 2.16×10^6 second-equivalents

43. In the scenario of Fig. Test II-8, what's the time-space value of d_t in meter-equivalents? Consider the speed of light and the length of the year to have the values stated in Question 47. As before, consider the earth's orbit as a perfect circle and the sun as a stationary point of reference.

A. 1.30×10^{15} meter-equivalents

B. 2.36×10^{15} meter-equivalents

C. 2.59×10^{15} meter-equivalents

D. 4.73×10^{15} meter-equivalents

E. 9.45×10^{15} meter-equivalents

44. The larger of the two definable angles between a line and a plane has a measure that can range anywhere between

    A. 0° and 360°.
    B. 90° and 180°.
    C. 180° and 270°.
    D. 270° and 360°.
    E. 180° and 360°.

45. If we double the lengths of all the edges of a perfect tesseract, its four-space hypervolume increases by a factor of

    A. 2.
    B. 4.
    C. 8.
    D. 16.
    E. 32.

46. Figure Test II-9 shows a hypothetical set of coordinates for Cartesian four-space. What, if any, problem exists with this rendition?

    A. It contains one too many lines (axes); we need, and should have, only three lines (axes) here.
    B. It can't uniquely portray points in four-space in our "real world," because we can't, in practice, make four lines intersect at a common point and remain mutually perpendicular.

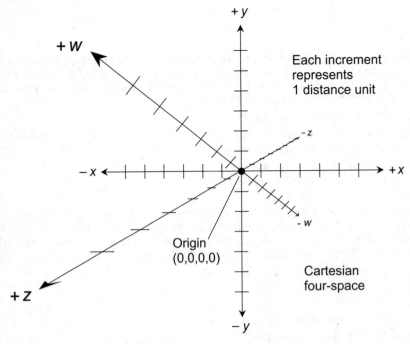

FIGURE TEST II-9 • Illustration for Part II Test Questions 46 and 47.

C. It does not contain enough axes; we must add an axis to represent time, so that we end up with five axes in total.

D. A four-space coordinate system cannot exist in Cartesian form; we must either graduate one of the axes in nonuniform increments, or else represent it as a curve.

E. No problem exists with this system.

47. **How can we change the coordinate system shown in Fig. Test II-9 so that it represents 4D time-space?**

A. We can ensure that the w, x, y, and z axes remain mutually perpendicular at the origin and then imagine (but not attempt to draw) time as an additional t axis, running from the past (negative values of t), through the present ($t = 0$), and toward the future (positive values of t).

B. We can change the x axis to a t axis to represent time, running from the past (negative values of t), through the present ($t = 0$), and toward the future (positive values of t), and leave everything else the same.

C. We can remove the w axis, ensure that the x, y, and z axes remain mutually perpendicular at the origin, and then imagine (but not attempt to draw) time as an additional t axis, running from the past (negative values of t), through the present ($t = 0$), and toward the future (positive values of t).

D. We can convert the entire system to a set of celestial coordinates that portrays values of right ascension, declination, azimuth, and elevation.

E. We can't.

48. **Imagine a unit circle in the Cartesian plane, and a ray that emanates from the origin (0,0) outward and upward toward the left, so that we have to turn precisely 45° clockwise to get from the negative x axis to the ray. What's the x-value of the point where the ray passes through the unit circle, accurate to three decimal places?**

A. 0.707
B. −0.500
C. 0.866
D. −0.866
E. −0.707

49. **In the situation of Question 48, what's the y-value of the point where the ray passes through the unit circle, accurate to three decimal places?**

A. 0.707
B. −0.500
C. 0.866
D. −0.866
E. −0.707

50. **One hour of right ascension, as an astronomer would define it, represents an angle equivalent to**

A. 1°.
B. 10°.
C. 15°.
D. 30°.
E. 60°.

Final Exam

Do not refer to the text when taking this test. You may draw diagrams or use a calculator if necessary. A good score is at least 75 correct. Answers are in the back of the book. It's best to have a friend check your score the first time, so you won't memorize the answers if you want to take the test again.

1. Each vertex of a triangle corresponds to a specific interior angle that measures
   A. more than 0 rad but less than π/2 rad.
   B. more than π/2 rad but less than π rad.
   C. more than π/2 rad but less than 2π rad.
   D. more than 0 rad but less than π rad.
   E. more than −π/2 rad but less than π/2 rad.

2. Which of the following statements is true?
   A. All trapezoids are squares.
   B. All trapezoids are rectangles.
   C. All rhombuses are squares.
   D. All squares are rectangles.
   E. All rectangles are rhombuses.

3. In a *convex* Euclidean plane polygon, the measure of each interior angle must remain less than
   A. 45°.
   B. 90°.
   C. 180°.
   D. 270°.
   E. 360°.

4. We say that two lines run parallel to each other if and only if they don't intersect anywhere, and also that they
   A. lie in the same plane.
   B. run perpendicular to each other.
   C. lie infinitely far apart.
   D. run askew relative to each other.
   E. have undefined separation distance.

5. Suppose that we encounter a Euclidean plane triangle whose sides measure exactly 23, 23, and 37 meters long. We've found
   A. a reflex triangle.
   B. a right triangle.
   C. an isosceles triangle.

D. an equilateral triangle.

E. None of the above

6. When we encounter a rectangle, we can have complete confidence that the measures of either pair of opposite interior angles add up to

A. π rad.

B. 2π rad.

C. $\pi/2$ rad.

D. $3\pi/2$ rad.

E. 3π rad.

7. Figure Exam-1 portrays two lines L and M that both intersect a transversal line N. All three lines L, M, and N lie in a single flat plane. Line N intersects line L at point P. Line N intersects line M at point Q. As a result, we get eight angles s through z, as shown. Suppose that we scrutinize all eight angles and find that angle x has a slightly larger measure than angle t. From this information, we can have absolute confidence that

A. lines L and M intersect somewhere.

B. angles v and y have equal measure.

C. angles w and s have equal measure.

D. angles u and z have equal measure.

E. lines L and M don't intersect anywhere.

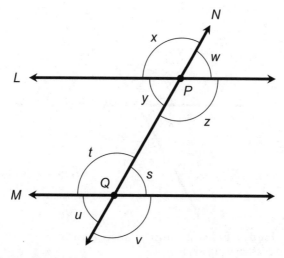

FIGURE EXAM-1 . Illustration for Final Exam Question 7.

8. In order to "qualify" as a true Euclidean plane quadrilateral, a geometric figure must have all of the following characteristics except one. Which one?

   A. Four vertices, all of which lie in the same plane

   B. Four sides, all of which have finite, positive, nonzero length

   C. Four interior angles whose measures add up to π rad

   D. Four interior angles, each of which has positive measure

   E. Four sides, all of which are straight line segments

9. Imagine a regular Euclidean plane polygon with interior angles that all measure 144°. What's the measure of each exterior angle?

   A. 36°

   B. 54°

   C. 216°

   D. 234°

   E. 324°

10. Figure Exam-2 portrays two lines L and M that intersect at point P. As a result, we get four angles w through z, as shown. We can have absolute confidence that

    A. angles w and x complement each other.

    B. angles x and y complement each other.

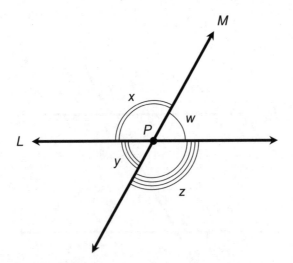

FIGURE EXAM-2 . Illustration for Final Exam Questions 10 and 11.

C. angles *w* and *y* complement each other.

D. More than one of the above

E. None of the above

11. **In the situation shown by Fig. Exam-2, we can have absolute confidence that**

A. angles *w* and *x* have equal measure.

B. angles *x* and *y* have equal measure.

C. angles *w* and *y* have equal measure.

D. More than one of the above

E. None of the above

12. Consider a specific line *L* in Euclidean three-space. Let *R* represent a point that does not lie on *L*. How many different lines can we find that pass through point *R* and run parallel to *L*?

A. None

B. One

C. Two

D. Three

E. Infinitely many

13. Consider a specific line *N* in Euclidean three-space. Let *X* represent a point that does not lie on *N*. How many different lines can we find that pass through point *X* and run askew to *N*?

A. None

B. One

C. Two

D. Three

E. Infinitely many

14. We can have absolute confidence that all four of the triangles shown in Fig. Exam-3 exhibit

A. direct similarity.

B. direct congruence.

C. inverse similarity.

D. inverse congruence.

E. More than one of the above

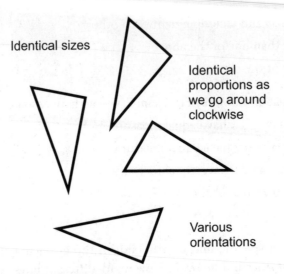

FIGURE EXAM-3 . Illustration for Final Exam
Question 14.

15. We can have absolute confidence that all four of the triangles shown in
Fig. Exam-4 exhibit

   A. direct similarity.

   B. direct congruence.

   C. inverse similarity.

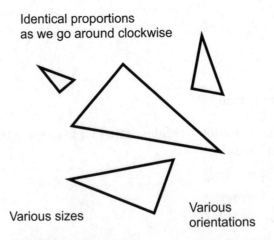

FIGURE EXAM-4 . Illustration for Final Exam
Question 15.

D. inverse congruence.

E. More than one of the above

16. In a Euclidean plane, the ratio of any circle's circumference to its diameter equals *precisely*

A. 22/7.

B. 3.14.

C. 3.14159.

D. the Arccosine of −1.

E. None of the above

17. If we have an ellipse with known dimensions and calculate its ellipticity, we get a number that tells us

A. the ratio of the interior area to the circumference.

B. the ratio of the circumference to the interior area.

C. the average of the lengths of the semiaxes.

D. how much the figure differs from a perfect circle.

E. how much the figure differs from an inscribed regular polygon.

18. Which of the following maneuvers constitutes "cheating" in a geometric construction with a compass, pencil, and straight edge?

A. Drawing a "random" arc with the compass, centered at an arbitrary point

B. Defining a specific point by making a dot with the pencil

C. Marking the straight edge to quantify the length of a line segment

D. Using the compass to draw a circle centered at the end of a line segment

E. Referencing a distance using the compass

19. In the situation shown by Fig. Exam-5, the area enclosed by $\triangle QPR$ equals

A. $(th + sh)/2$.

B. $thu + shv$.

C. $uv + ts$.

D. $2hu + 2hv$.

E. $[(t + s)uv]/2$.

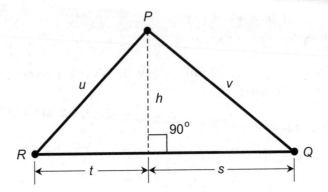

FIGURE EXAM-5 . Illustration for Final Exam Question 19.

20. Imagine a circle that lies in a Euclidean plane, and that has a radius of exactly 1 meter. Suppose that we circumscribe this circle with a regular polygon having n sides, and then we increase n without limit, all the while making sure that the polygon fits "tightly" around the circle. As we carry out this process, the interior area of the polygon approaches

A. π meters.

B. the square root of 2 meters.

C. the square root of π meters.

D. $\pi/2$ meters.

E. 1 meter.

21. In the situation shown by Fig. Exam-6, suppose that line segment ST runs parallel to line segment RQ. In that case, we know that the Euclidean plane quadrilateral $STQR$ is a

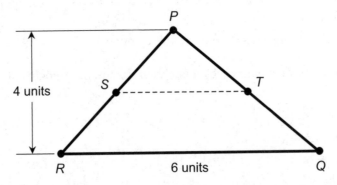

FIGURE EXAM-6 . Illustration for Final Exam Questions 21 through 23.

A. parallelogram.

B. rhombus.

C. trapezoid.

D. reflex figure.

E. truncated rectangle.

22. In the situation of Fig. Exam-6, suppose that line segment *ST* not only runs parallel to line segment *RQ*, but lies exactly 2 units from line segment *RQ*. How long is line segment *ST*?

   A. We need more information to calculate it.

   B. 3 units

   C. The square root of 6 units

   D. 2 units

   E. The square root of 10 units

23. In the situation of Fig. Exam-6, suppose that line segment *ST* runs parallel to line segment *RQ* and lies exactly 2 units from line segment *RQ*. What's the area enclosed by quadrilateral *STQR*?

   A. We need more information to calculate it.

   B. 6 square units

   C. The square root of 24 square units

   D. 9 square units

   E. The square root of 48 square units

24. How can we use a compass, pencil, and straight edge to construct an angle whose measure equals 22.5°?

   A. We can draw a line segment, construct its perpendicular bisector, bisect the resulting right angle to get 45°, and then bisect the 45° angle to get 22.5°.

   B. We can construct a rhombus and then bisect one of its vertex angles.

   C. We can construct an equilateral triangle and then bisect one of the vertex angles.

   D. We can construct a regular hexagon and then bisect one of its vertex angles.

   E. We can't, unless we "cheat" and use a calibrated compass.

25. Suppose that you want to construct a rhombus with a compass and straight edge. What should you do first?

A. Construct two intersecting arcs.

B. Construct two parallel lines.

C. Construct two perpendicular lines.

D. Construct an equilateral triangle.

E. Construct a square.

26. Imagine a trapezoid defined by points P, Q, R, and S, which we encounter in that order as we go clockwise around the figure. Imagine that the sides have lengths d, e, f, and g as shown in Fig. Exam-7. Let d represent the base length, let h represent the height (vertical dashed line), let x represent the angle between the sides having length d and e, and let y represent the angle between the sides having lengths g and d. Suppose that the sides having lengths d and f (line segments RS and PQ) are parallel. Let m represent the length of the median of the trapezoid as shown by the horizontal dashed line. Which of the following equations holds true in **all possible situations** of the sort portrayed by this generic drawing?

A. $x = y$

B. $m = (d + f)/2$

C. $e = g$

D. $h < m$

E. $h = m$

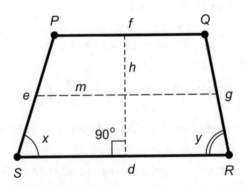

FIGURE EXAM-7 . Illustration for Final Exam Questions 26 through 28.

27. Which of the following equations defines the perimeter B of trapezoid *PQRS* as shown in Fig. Exam-7? (The 1/2 power denotes the positive square root.)

    A. $B = h + m$

    B. $B = 2h + 2m$

    C. $B = d + e + f + g$

    D. $B = (defg)^{1/2}$

    E. $B = mh$

28. Which of the following equations defines the area A enclosed by trapezoid *PQRS* as shown in Fig. Exam-7? (The 1/2 power denotes the positive square root.)

    A. $A = h + m$

    B. $A = 2h + 2m$

    C. $A = d + e + f + g$

    D. $A = (defg)^{1/2}$

    E. $A = mh$

29. Consider a circle represented by the following equation in Cartesian coordinates:

$$(x - 4)^2 + (y + 1)^2 = 64$$

    What are the coordinates of the circle's center?

    A. $(4, -1)$

    B. $(-4, 1)$

    C. $(-1, 4)$

    D. $(1, -4)$

    E. We need more information to figure it out.

30. What's the radius of the circle with the equation described in Question 29?

    A. 64 units

    B. 32 units

    C. 16 units

    D. 8 units

    E. We need more information to figure it out.

FIGURE EXAM-8 . Illustration for Final
Exam Question 31.

31. Figure Exam-8 illustrates an exterior angle θ for a "generic" Euclidean
plane polygon. Which of the following inequalities describes the range of
values, in radians, that θ can have?

   A. $0 < \theta < \pi$

   B. $0 < \theta < \pi/2$

   C. $0 < \theta < \pi/4$

   D. $\pi/2 < \theta < \pi$

   E. $-\pi/2 < \theta < \pi/2$

32. In Euclidean three-space, how many different lines can run *perpendicular*
to a given plane through a specific point that does not lie in that plane?

   A. None

   B. One

   C. Two

   D. Three

   E. Infinitely many

33. In Euclidean three-space, how many different lines can run *parallel* to a
given plane through a specific point that does not lie in that plane?

   A. None

   B. One

   C. Two

   D. three

   E. Infinitely many

34. **The ideal straight edge for carrying out a geometric construction**

    A. has an angle reference scale, preferably calibrated in degrees.

    B. is calibrated for distance, preferably in metric units such as millimeters.

    C. has little holes in it to make small circles or to define points.

    D. is a drafting triangle with two 45° angles and one 90° angle.

    E. is an uncalibrated, flat object with at least one straight side.

35. **If we want to define a specific straight line in Euclidean geometry, we must precisely know the**

    A. location of one point.

    B. locations of two points.

    C. locations of three points.

    D. locations and orientations of three planes.

    E. locations and orientations of four planes.

36. **Figure Exam-9 illustrates a "generic" regular Euclidean plane polygon. It has n sides, each of length s units. Each interior angle measures θ radians.**

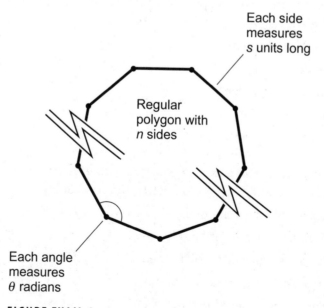

Each side measures s units long

Regular polygon with n sides

Each angle measures θ radians

FIGURE EXAM-9 . Illustration for Final Exam Questions 36 and 37.

Suppose that we let *n* increase without limit, but we also make sure that the total perimeter of the polygon remains constant. What happens to *s* as we do this?

A. It approaches 0.

B. It approaches 1 divided by the perimeter of the polygon.

C. It approaches π divided by the perimeter of the polygon.

D. It approaches the square root of π divided by the perimeter of the polygon.

E. We can't say unless we use calculus to figure it out.

37. In the situation shown by Fig. Exam-9 and described in Question 36, what happens to the value of *θ* as we increase *n* without limit?

A. It approaches 0 rad.

B. It approaches π/4 rad.

C. It approaches π/2 rad.

D. It approaches π rad.

E. It approaches 2π rad.

38. Imagine the set of all possible isosceles triangles in a specific Euclidean plane. We can have complete confidence that if we choose any two of these triangles "at random," they'll turn out

A. directly congruent.

B. directly similar.

C. inversely congruent.

D. inversely similar.

E. None of the above

39. If we can identify *at least one point* that two *different* planes in Euclidean three-space share, then we know that the two planes

A. run parallel to each other.

B. intersect in a straight line.

C. run perpendicular to each other.

D. run askew relative to each other.

E. intersect in a pair of parallel lines.

40. If two *different* planes in Euclidean three-space share *no points whatso-ever*, then we know that the two planes

A. run parallel to each other.

B. intersect in a straight line.

C. run perpendicular to each other.

D. run askew relative to each other.

E. intersect in a pair of parallel lines.

41. Carefully inspect Fig. Exam-10. Suppose that you have a line segment containing a point P, as shown in Fig. Exam-10A. You set your drafting compass for a moderate span and construct two arcs opposite each other, both centered at P and intersecting the line segment at points Q and R (Fig. Exam-10B). Next, you roughly double the span of the compass and then construct an arc centered at Q and another arc centered at R, so that the two arcs have the same radius and intersect each other at some point away from the line segment (Fig. Exam-10C).

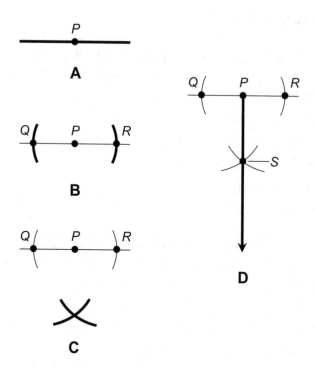

FIGURE EXAM-10 . Illustration for Final Exam Question 41.

Finally, you use your straight edge to draw a ray that originates at *P*, and that passes through the intersection point *S* of the two arcs you just made (Fig. Exam-10D). In this situation, you can have absolute confidence that

A. line segment *PS* has the same length as line segment *QR*.

B. ray *PS* runs perpendicular to line segment *QR*.

C. line segment *QP* has the same length as line segment *PR*.

D. More than one of the above

E. None of the above

42. In order for two lines in Euclidean three-space to run askew relative each other, they must *not*

A. intersect at any point.

B. lie in the same plane.

C. run parallel to each other.

D. define a pair of vertical angles.

E. All of the above

43. If three lines all share *exactly one* point in Euclidean three-space, which of the following statements can we make, with absolute certainty, about those lines?

A. They all lie in the same plane

B. They're all parallel to each other

C. They're all askew relative to each other

D. They all coincide

E. None of the above

44. Figure Exam-11 shows two points on the Cartesian coordinate system. What are the coordinates of point *P*, expressed as an ordered pair of the form (*x,y*)?

A. (4,–5)

B. (–5,4)

C. (–4,5)

D. (5,–4)

E. (–4,–5)

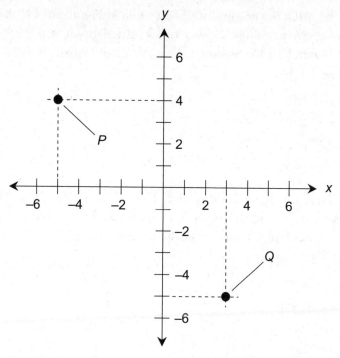

FIGURE EXAM-11 . Illustration for Final Exam Questions 44 through 48.

45. What are the coordinates of point Q in Fig. Exam-11, expressed as an ordered pair of the form (x,y)?

A. (3,5)

B. (−3,5)

C. (5,3)

D. (−5,3)

E. (3,−5)

46. How far from the origin does point P lie in Fig. Exam-11? Assume that the coordinate values, as you've identified them, are mathematically exact. Round off the answer to three decimal places.

A. 4.472 units

B. 4.500 units

C. 6.000 units

D. 6.403 units

E. 6.667 units

47. How far from the origin does point Q lie in Fig. Exam-11? Assume that the coordinate values, as you've identified them, are mathematically exact. Round off the answer to three decimal places.

    A. 6.000 units

    B. 5.831 units

    C. 5.657 units

    D. 5.333 units

    E. 5.111 units

48. How far from each other do points *P* and Q lie in Fig. Exam-11? Assume that the coordinate values, as you've identified them, are mathematically exact. Round off the answer to three decimal places.

    A. 12.234 units

    B. 12.042 units

    C. 11.000 units

    D. 10.000 units

    E. 9.667 units

49. A tetrahedron has

    A. four vertices, four edges, and four faces.

    B. four vertices, six edges, and four faces.

    C. six vertices, six edges, and four faces.

    D. six vertices, eight edges, and four faces.

    E. eight vertices, eight edges, and four faces.

50. In a *regular* tetrahedron, each face constitutes

    A. an equilateral triangle.

    B. a right triangle.

    C. an obtuse triangle.

    D. a reflex triangle.

    E. a square.

51. If we increase the volume of a cube by a factor of 125, then its surface area increases by a factor of

    A. 5.

    B. the square root of 50.

C. 10.

D. the square root of 125.

E. 25.

52. Suppose that we encounter a Euclidean plane triangle whose sides measure exactly 20, 48, and 52 meters long. We've found

A. an acute triangle.

B. a right triangle.

C. an isosceles triangle.

D. an equilateral triangle.

E. an obtuse triangle.

53. Figure Exam-12 shows the graphs of three equations in Cartesian coordinates. The graphs appear as a parabola (A), a circle (B), and a straight line (C). Let's call the corresponding equations "Equation A," "Equation B," and "Equation C," even though we don't know any numerical specifics

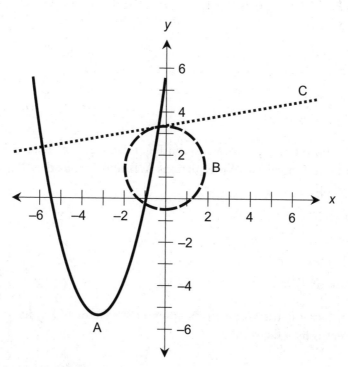

FIGURE EXAM-12 . Illustration for Final Exam Questions 53 through 56.

about them. Based on the visual information in Fig. Exam-12, how many distinct real-number solutions exist for Equations A and B as a pair?

A. None

B. One

C. Two

D. Three

E. Infinitely many

54. Based on the visual information in Fig. Exam-12, how many distinct real-number solutions exist for Equations B and C as a pair?

A. None

B. One

C. Two

D. Three

E. Infinitely many

55. Based on the visual information in Fig. Exam-12, how many distinct real-number solutions exist for Equations A and C as a pair?

A. None

B. One

C. Two

D. Three

E. Infinitely many

56. Based on the visual information in Fig. Exam-12, how many distinct real-number solutions exist for Equations A, B, and C considered all together?

A. None

B. One

C. Two

D. Three

E. Infinitely many

57. If we increase the volume of a sphere by a factor of 64, its surface area increases by a factor of

A. 4.

B. 8.

C. 16.

D. the square root of 128.

E. the cube root of 2048.

58. **What's the magnitude of the vector 2i + 2j − 2k in Cartesian three-space?**

A. 4

B. The square root of 6

C. 8

D. The square root of 8

E. The square root of 12

59. **Figure Exam-13 shows three plane regions with different boundary definitions. Based on the assumption that the figures all have the same general size and shape, which of the following statements holds true?**

A. The regions all have identical interior areas, and they all have identical perimeters.

B. Region A has greater interior area than region B, which in turn has greater interior area than region C; however, all three regions have identical perimeters.

C. Region A has greater perimeter than region B, which in turn has greater perimeter than region C; however, all three regions have identical interior areas.

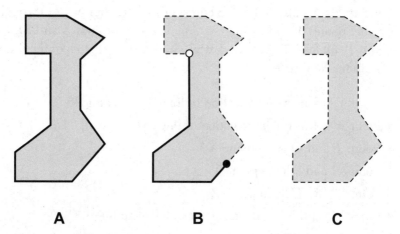

A   **B**   **C**

FIGURE EXAM-13 . Illustration for Final Exam Question 59.

D. Region A has greater perimeter than region B, which in turn has greater perimeter than region C. In addition, region A has greater interior area than region B, which in turn has greater interior area than region C.

E. We cannot define the interior areas or perimeters of any of the figures shown here, because the boundary specifications aren't clear.

60. In Cartesian time-space, each point follows its own time line. Assuming that no point moves with respect to the origin, all the points follow time lines that run

A. "parallel" to all the other time lines and "perpendicular" to three-space.

B. "parallel" to all the other time lines and "parallel" to three-space.

C. "perpendicular" to all the other time lines and "perpendicular" to three-space.

D. "perpendicular" to all the other time lines and "parallel" to three-space.

E. "on the surface" of a light cone with the origin at its apex.

61. Imagine all possible right triangles in a Euclidean plane. If we choose any two of them "at random," we can have absolute confidence that they'll both

A. be inversely similar.

B. be directly similar.

C. have equal perimeters.

D. have equal interior areas.

E. conform to the theorem of Pythagoras.

62. Refer to Fig. Exam-14. Let X represent a plane that passes through two parallel planes Y and Z, intersecting Y and Z in lines L and M. Define points P, Q, R, S, T, U, V, and W as shown, such that all of the following conditions hold true:

• Point V lies at the intersection of lines L, PQ, and RS

• Point W lies at the intersection of lines M, PQ, and TU

• Points P and Q lie in plane X

• Points R and S lie in plane Y

• Points T and U lie in plane Z

• Lines PQ and RS both run perpendicular to line L

• Lines PQ and TU both run perpendicular to line M

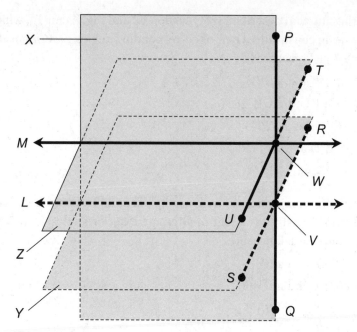

FIGURE EXAM-14 . Illustration for Final Exam Questions 62 through 66.

Which of the following constitutes a pair of vertical angles, assuring us that they have equal measure?

A. ∠TWP and ∠RVW

B. ∠TWP and ∠UWQ

C. ∠QWT and ∠PVS

D. ∠PWU and ∠QVR

E. ∠PVS and ∠WVR

63. In the situation described by Question 62 and Fig. Exam-14, which of the following constitutes a pair of alternate interior angles, assuring us that they have equal measure?

A. ∠TWP and ∠RVW

B. ∠TWP and ∠UWQ

C. ∠QWT and ∠PVS

D. ∠PWU and ∠QVR

E. ∠PVS and ∠WVR

64. **In the situation described by Question 62 and Fig. Exam-14, which of the following constitutes a pair of corresponding angles, assuring us that they have equal measure?**

    A. $\angle TWP$ and $\angle RVW$

    B. $\angle TWP$ and $\angle UWQ$

    C. $\angle QWT$ and $\angle PVS$

    D. $\angle PWU$ and $\angle QVR$

    E. $\angle PVS$ and $\angle WVR$

65. **In the situation described by Question 62 and Fig. Exam-14, which of the following constitutes a pair of alternate exterior angles, assuring us that they have equal measure?**

    A. $\angle TWP$ and $\angle RVW$

    B. $\angle TWP$ and $\angle UWQ$

    C. $\angle QWT$ and $\angle PVS$

    D. $\angle PWU$ and $\angle QVR$

    E. $\angle PVS$ and $\angle WVR$

66. **In the situation described by Question 62 and Fig. Exam-14, which of the following constitutes a pair of adjacent angles, assuring us that they're supplementary?**

    A. $\angle TWP$ and $\angle RVW$

    B. $\angle TWP$ and $\angle UWQ$

    C. $\angle QWT$ and $\angle PVS$

    D. $\angle PWU$ and $\angle QVR$

    E. $\angle PVS$ and $\angle WVR$

67. **Which of the following geometric objects represents a true mathematical function when we work with it in a polar coordinate plane, but not when we work with it in a Cartesian coordinate plane?**

    A. A circle centered at the origin

    B. A straight, horizontal line that passes through the origin

    C. A straight, horizontal line that does not pass through the origin

    D. A parabola that opens upward and whose vertex lies at the origin

    E. A parabola that opens upward and whose vertex does not lie at the origin

68. Imagine two rays emanating outward from the center point of a circle in a Euclidean plane. Each of the two rays intersects the circle at a point; call these points P and Q. Suppose that the distance between P and Q, as expressed along the arc of the circle, equals the radius of the circle. In this scenario, the measure of the angle between the rays equals

   A. 1°.

   B. 30°.

   C. 45°.

   D. 60°.

   E. None of the above

69. What's the total surface area of the rectangular prism shown in Fig. Exam-15?

   A. 763 square inches

   B. 1144 square inches

   C. 1526 square inches

   D. 3052 square inches

   E. We need more information to calculate it.

70. What's the volume of the rectangular prism shown in Fig. Exam-15?

   A. 49 cubic inches

   B. 98 cubic inches

   C. 2401 cubic inches

   D. 3795 cubic inches

   E. We need more information to calculate it.

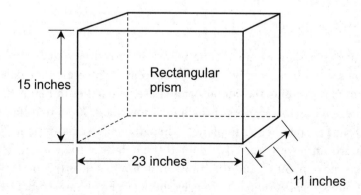

FIGURE EXAM-15 . Illustration for Final Exam Questions 69 and 70.

71. In polar coordinates, the equation $4r = 5\theta$ represents a
    A. straight line.
    B. circle.
    C. spiral.
    D. three-leafed rose.
    E. hyperbola.

72. If we quadruple the length of one semiaxis of an ellipsoid while leaving the other two semiaxes unchanged, the volume of the enclosed solid increases by a factor of
    A. 2.
    B. the square root of 8.
    C. the cube root of 32.
    D. 4.
    E. 8.

73. Suppose that we want to uniquely define a geometric plane in Cartesian three-space. We can accomplish this task if we can determine
    A. the coordinates of one point in the plane and the direction of a vector that runs parallel to the plane.
    B. the coordinates of two points in the plane and the direction of a vector that runs parallel to the plane.
    C. the coordinates of one point in the plane and the direction of a vector that runs normal to the plane.
    D. the back-end points of two vectors that both run normal to the plane.
    E. Any of the above

74. Consider a slant circular cone whose base radius equals r and height equals h, as shown in Fig. Exam-16A. Point P represents the cone's apex. Point C represents the center of the base, which lies in plane X. Point Q represents the projection of the apex onto plane X, so that line segment PQ runs perpendicular to plane X. Imagine that we move point P straight upward until we've exactly doubled the height of the cone to $2h$, but we don't move point C, and the transformation has no effect on the location of point Q. We get a taller cone, as shown in Fig. Exam-16 B. How do the volumes of these two cones compare?

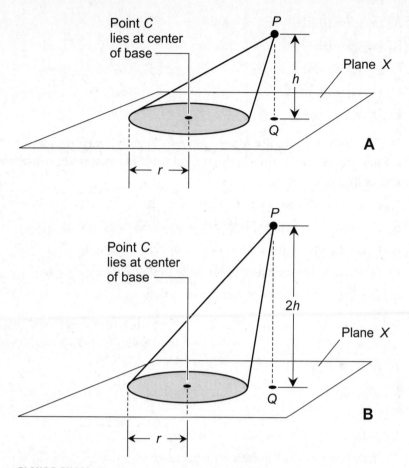

FIGURE EXAM-16 . Illustration for Final Exam Question 74.

A. The taller cone has twice the volume of the shorter cone.

B. The taller cone has the square root of 8 times the volume of the shorter cone.

C. The taller cone has the cube root of 16 times the volume of the shorter cone.

D. The taller cone has four times the volume of the shorter cone.

E. We need more information to answer this.

75. Consider the point $(\theta_0, r_0) = (3\pi/2, 16)$ in mathematician's polar coordinates. What's the ordered-pair (x_0, y_0) representation of this point in Cartesian coordinates?

A. $(x_0, y_0) = (0, -16)$

B. $(x_0, y_0) = (0, -4)$

C. $(x_0, y_0) = (4, 0)$

D. $(x_0, y_0) = (4, 16)$

E. $(x_0, y_0) = (-4, -4)$

76. With a drafting compass *alone*, you can "legally" perform all of the following actions, according to the formal rules for geometric construction, *except one*. Which one?

A. Draw a circle centered at a defined point.

B. Replicate the distance between any two defined points.

C. Draw an arc centered at a "randomly" chosen point.

D. Determine the measure of an angle in degrees.

E. Draw a circle whose center lies on a defined line.

77. What's the sum of the vectors $(1, -5, 6)$ and $(0, 7, -12)$ in Cartesian three-space?

A. $(0, -35, -72)$

B. $(1, -12, 6)$

C. $(1, 2, -6)$

D. $(-1, 12, -18)$

E. We need more information to calculate it.

78. From the information shown in Figure Exam-17, we can deduce the fact that

A. the tangent of $90°$ is undefined.

B. the tangent of $45°$ equals 1.

C. the tangent of $135°$ equals -1.

D. the tangent of $315°$ equals -1.

E. All of the above

79. Based on the notion that the speed of light in free space equals 300,000 kilometers per second, we can define 1 minute of time as

A. 5000 kilometer-equivalents.

B. 18,000,000 kilometer-equivalents.

C. 300,000 cubic kilometers.

D. 60 cubic seconds.

E. None of the above

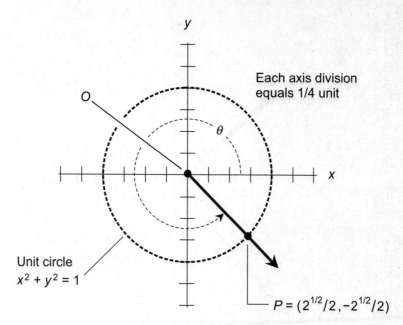

Each axis division equals 1/4 unit

Unit circle
$x^2 + y^2 = 1$

$P = (2^{1/2}/2, -2^{1/2}/2)$

FIGURE EXAM-17 . Illustration for Final Exam Question 78.

80. Imagine a rectangular prism that measures exactly 100 meters high, 200 meters wide, and 400 meters deep. Suppose that it forms from nothing, exists in free space for exactly 0.01 second, and then vanishes. What's its hypervolume in quartic kilometer-equivalents? Assume that the free-space speed of light equals exactly 3×10^5 kilometers per second.

    A. 6 quartic kilometer-equivalents

    B. 12 quartic kilometer-equivalents

    C. 18 quartic kilometer-equivalents

    D. 24 quartic kilometer-equivalents

    E. We need more information to calculate it.

81. What's the dot product of the two vectors shown in Fig. Exam-18?

    A. $\mathbf{c} \bullet \mathbf{d} = 126$

    B. $\mathbf{c} \bullet \mathbf{d} = 23$

    C. $\mathbf{c} \bullet \mathbf{d} = 35$

    D. $\mathbf{c} \bullet \mathbf{d} = 0$

    E. We need more information to figure it out.

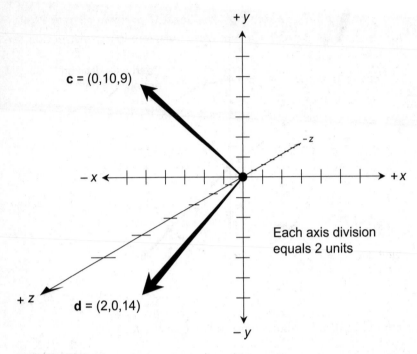

$c = (0,10,9)$

$d = (2,0,14)$

Each axis division
equals 2 units

FIGURE EXAM-18 . Illustration for Final Exam Question 81.

82. What's the distance between the origin and the point $(1,1,1,1,1,1,1,1,1,1)$ in Cartesian 10-space? Assume the coordinate values to be exact.

A. 10 units

B. 100 units

C. The square root of 10 units

D. The 10th root of 10 units

E. 1 unit

83. What's the distance between the origin and the point $(2,2,2,2,2,2,2,2,2,2)$ in Cartesian 10-space? Assume the coordinate values to be exact.

A. The square root of 40 units

B. 200 units

C. The 10th root of 20 units

D. The square root of 20 units

E. 2 units

84. With a straight edge *alone,* you can "legally" perform all of the following actions, according to the formal rules for geometric construction, *except one.* Which one?

   A. Draw a ray that starts at a defined point.

   B. Draw a line segment connecting two known points.

   C. Duplicate a line segment.

   D. Construct two lines that intersect at a single point.

   E. Draw a line that intersects a defined circle at two "random" points.

85. What's the maximum number of dimensions that can theoretically exist in Cartesian hyperspace?

   A. Three

   B. Four

   C. Five

   D. It depends on whether or not we include time.

   E. No maximum exists!

86. Figure Exam-19 is a polar-coordinate graph showing a particular point *P.* Each radial division (where radial divisions show up as concentric circles)

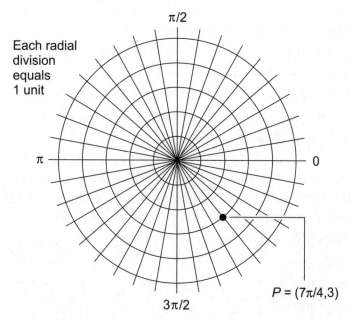

FIGURE EXAM-19 . Illustration for Final Exam Questions 86 and 87.

represents 1 unit. Based on this information, what's the x coordinate of P in the Cartesian xy-plane?

A. $3^{1/2}$

B. $6^{1/2}$

C. $(9/2)^{1/2}$

D. $7\pi/4$

E. $-(9/2)^{1/2}$

87. In the situation of Fig. Exam-19, what's the y coordinate of point P in the Cartesian xy-plane?

A. $3^{1/2}$

B. $-6^{1/2}$

C. $(9/2)^{1/2}$

D. $7\pi/4$

E. $-(9/2)^{1/2}$

88. Suppose that we want to determine the equation of a geometric line in a Cartesian three-space coordinate system. We can accomplish this task if we can find

A. the coordinates of one point on the line and the direction numbers for a vector that runs parallel to the line.

B. the coordinates of one point on the line and the direction numbers for a vector that runs normal to the line.

C. the direction numbers for two vectors that run normal to the line.

D. the direction numbers for two vectors that run parallel to the line.

E. Any of the above

89. Figure Exam-20 illustrates a set of three-space coordinates commonly used by astronomers. What does the angular dimension θ, expressed in degrees, represent here?

A. Celestial longitude

B. Declination

C. Azimuth

D. Right ascension

E. Elevation

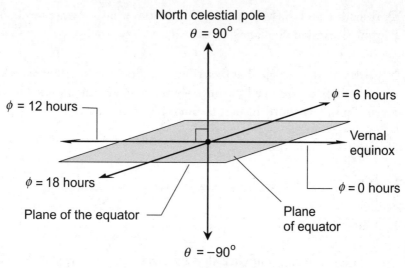

North celestial pole
$\theta = 90°$

$\phi = 6$ hours

$\phi = 12$ hours

Vernal
equinox

$\phi = 18$ hours

Plane of the equator

$\phi = 0$ hours

Plane
of equator

$\theta = -90°$

South celestial pole

FIGURE EXAM-20 . Illustration for Final Exam Questions 89 and 90.

90. **What does the angular dimension ϕ, expressed in hours, represent in the coordinate system of Fig. Exam-20?**

A. Celestial longitude

B. Declination

C. Azimuth

D. Right ascension

E. Elevation

91. **What's the sum of the vectors (2,5) and (−7,−10) in Cartesian two-space?**

A. (−5,−5)

B. (5,−15)

C. (9,−5)

D. (−8,−2)

E. (−14,−50)

92. **What's the dot product of the vectors (2,5) and (−7,−10) in Cartesian two-space?**

A. −64

B. −10

C. 4

D. 6

E. 36

93. In Cartesian time-space that describes our "real world," how many coordinate values do we need to uniquely name or define a point that lasts for an "infinitely short" instant in time?

A. Two

B. Three

C. Four

D. Five

E. Infinitely many

94. We can have total confidence that two triangles are inversely similar if they exhibit

A. inverse congruence.

B. direct similarity and the same orientation.

C. direct similarity and the same size.

D. direct congruence and different orientations.

E. Any of the above

95. In so-called navigator's polar coordinates, we don't allow the range to have negative values. Why?

A. It results in an undefined quotient, rendering it impossible to define the position of a point.

B. It produces relations but not always true mathematical functions.

C. It requires us to define angular values going clockwise, when we should always define them going counterclockwise.

D. In the "real world," nothing can lie any closer to us than the point representing our own location.

E. All of the above

96. Imagine that S, T, and U represent three collinear points (they all fall along a single straight line), such that T lies between S and U. Which of the following four distance equations, if any, is false?

A. $ST + TU = SU$

B. $SU - ST = TU$

C. $SU - TU = ST$

D. $ST - TU = SU$

E. All of the above equations are true.

97. **Which of the following actions violates the formal rules for geometric construction?**

    A. Define the length of a line segment by laying the nonmarking tip of a compass at one end point and the marking tip at the other end point.

    B. Represent a line by running a pencil's tip along a straight edge for some distance, and then draw arrows at each end of the pencil mark.

    C. Create a "random" angle by using a straight edge to draw two rays that intersect at their back-end points.

    D. Construct a "random" circle with a compass set to any desired span.

    E. Duplicate a line segment over and over, endlessly (in your imagination), to create an infinitely complex Euclidean plane polygon.

98. **Imagine two distinct points P and Q on a non-Euclidean surface. The shortest possible path between P and Q that lies entirely on the surface is known as a**

    A. Riemannian curve.

    B. Lobachevskian curve.

    C. longitudinal curve.

    D. latitudinal curve.

    E. None of the above

99. **Figure Exam-21 illustrates a non-Euclidean 2D surface containing an irregular polygon with five sides, all of which are geodesic arcs. The surface in this illustration has**

    A. negative curvature.

    B. positive curvature.

    C. nongeodesic curvature.

    D. relativistic curvature.

    E. elliptical curvature.

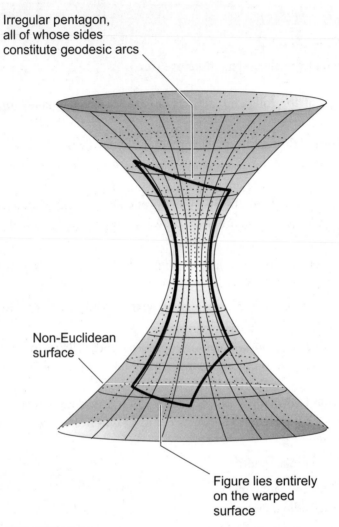

Irregular pentagon,
all of whose sides
constitute geodesic arcs

Non-Euclidean
surface

Figure lies entirely
on the warped
surface

FIGURE EXAM-21 . Illustration for Final Exam Questions 99 and 100.

100. If we take Fig. Exam-21 as a literal portrayal, we can have complete confidence that the measures of the interior angles of the irregular pentagon
sum up to

A. something more than 540°.

B. exactly 540°.

C. something less than 540°.

D. something more than 600°.

E. something more than 720°.

Answers to Quizzes, Tests, and Final Exam

Chapter 1	Chapter 3	Chapter 5	Test: Part I
1. B	1. D	1. C	1. E
2. C	2. B	2. D	2. B
3. A	3. A	3. C	3. C
4. B	4. C	4. A	4. B
5. B	5. D	5. B	5. A
6. C	6. A	6. A	6. D
7. A	7. A	7. D	7. D
8. B	8. C	8. C	8. D
9. D	9. D	9. C	9. C
10. D	10. B	10. A	10. C
			11. D
Chapter 2	Chapter 4	Chapter 6	12. B
1. D	1. B	1. D	13. A
2. B	2. B	2. C	14. E
3. D	3. D	3. C	15. B
4. C	4. C	4. B	16. C
5. C	5. D	5. A	17. D
6. B	6. B	6. B	18. B
7. D	7. D	7. C	19. E
8. B	8. C	8. D	20. C
9. A	9. C	9. A	21. E
10. C	10. A	10. B	22. B

23. D
24. E
25. E
26. A
27. C
28. D
29. D
30. C
31. E
32. D
33. A
34. D
35. D
36. E
37. C
38. A
39. A
40. E
41. D
42. A
43. B
44. C
45. D
46. A
47. B
48. A
49. E
50. E

Chapter 7
1. D
2. A
3. B
4. A
5. C
6. B
7. D
8. C

9. D
10. A

Chapter 8
1. B
2. B
3. C
4. C
5. D
6. A
7. C
8. C
9. C
10. B

Chapter 9
1. A
2. D
3. B
4. B
5. A
6. C
7. A
8. B
9. C
10. D

Chapter 10
1. C
2. C
3. A
4. D
5. B
6. B
7. C
8. A
9. D
10. A

Chapter 11
1. D
2. A
3. B
4. C
5. B
6. C
7. A
8. D
9. C
10. D

Test: Part II
1. C
2. C
3. E
4. A
5. D
6. E
7. A
8. E
9. D
10. B
11. A
12. E
13. B
14. A
15. B
16. A
17. D
18. C
19. C
20. E
21. D
22. D
23. A
24. A
25. B

26. A
27. D
28. B
29. A
30. A
31. A
32. C
33. A
34. B
35. D
36. E
37. D
38. B
39. A
40. C
41. D
42. B
43. E
44. B
45. D
46. B
47. C
48. E
49. A
50. C

Final Exam
1. D
2. D
3. C
4. A
5. C
6. A
7. A
8. C
9. A
10. E
11. C

12. B	34. E	56. B	78. D
13. E	35. B	57. C	79. B
14. E	36. A	58. E	80. D
15. A	37. D	59. A	81. A
16. D	38. E	60. A	82. C
17. D	39. B	61. E	83. A
18. C	40. A	62. B	84. C
19. A	41. D	63. C	85. E
20. A	42. E	64. A	86. C
21. C	43. E	65. D	87. E
22. B	44. B	66. E	88. A
23. D	45. E	67. A	89. B
24. A	46. D	68. E	90. D
25. B	47. B	69. C	91. A
26. B	48. B	70. D	92. A
27. C	49. B	71. C	93. C
28. E	50. A	72. D	94. A
29. A	51. E	73. C	95. D
30. D	52. B	74. A	96. D
31. A	53. C	75. A	97. E
32. B	54. B	76. D	98. E
33. E	55. C	77. C	99. A
			100. C

Suggested Additional Reading

Bluman, Alan, *Pre-Algebra Demystified*, 2nd ed. McGraw-Hill, 2011.

Gibilisco, Stan, *Algebra Know-It-All*. McGraw-Hill, 2008.

Gibilisco, Stan, *Mastering Technical Mathematics*, 3rd ed. McGraw-Hill, 2007.

Gibilisco, Stan, *Math Proofs Demystified*. McGraw-Hill, 2005.

Gibilisco, Stan, *Pre-Calculus Know-It-All*. McGraw-Hill, 2010.

Gibilisco, Stan, *Technical Math Demystified*. McGraw-Hill, 2006.

Gibilisco, Stan, *Trigonometry Demystified*. McGraw-Hill, 2003.

Huettenmueller, Rhonda, *Algebra Demystified*, 2nd ed. McGraw-Hill, 2011.

Kelley, W. Michael, *The Humongous Book of Geometry Problems*. Alpha, 2009.

Leff, Lawrence S., *Geometry the Easy Way*, 4th ed. Barron's Educational Series, 2009.

Long, Lynnette, *Painless Geometry*, 2nd ed. Barron's Educational Series, 2009.

Mlodinow, Leonard, *Euclid's Window*. Free Press, 2002.

Prindle, Anthony and Katie, *Math the Easy Way*, 2nd ed. Barron's Educational Series, 2009.

Rich, Barnett and Thomas, Christopher, *Schaum's Outline of Geometry*, 4th ed. McGraw-Hill, 2008.

Ryan, Mark, *Geometry for Dummies*, 2nd ed. Wiley, 2008.

Index

DeMYSTiFieD®

Hard stuff made easy

The DeMYSTiFieD series helps students master complex and difficult subjects. Each book is filled with chapter quizzes, final exams, and user friendly content. Whether you want to master Spanish or get an A in Chemistry, DeMYSTiFieD will untangle confusing subjects, and make the hard stuff understandable.

PRE-ALGEBRA DeMYSTiFied, 2e
Allan G. Bluman
ISBN-13: 978-0-07-174252-8 • $20.00

ALGEBRA DeMYSTiFied, 2e
Rhonda Huettenmueller
ISBN-13: 978-0-07-174361-7 • $20.00

CALCULUS DeMYSTiFied, 2e
Steven G. Krantz
ISBN-13: 978-0-07-174363-1 • $20.00

PHYSICS DeMYSTiFied, 2e
Stan Gibilisco
ISBN-13: 978-0-07-174450-8 • $20.00

Learn more. Do more.